FLUORIDE REMOVAL FROM GROUNDWATER BY ADSORPTION TECHNOLOGY

The occurrence, adsorbent synthesis, regeneration and disposal

ABDULAI SALIFU

FLUORIDE REMOVAL FROM GROUNDWATER BY ADSORPTION TECHNOLOGY

The occurrence, adsorbent synthesis, regeneration and disposal

DISSERTATION

Submitted in fulfillment of the requirements of

the Board for Doctorates of Delft University of Technology

and

of the Academic Board of the UNESCO-IHE

Institute for Water Education

for

the Degree of DOCTOR

to be defended in public on

Wednesday 4 October, 2017, at 15:00 hours

in Delft, the Netherlands

by

Abdulai SALIFU

Master of Science in Environmental Engineering

University of Newcastle Upon Tyne, England

born in Kumasi, Ghana

This dissertation has been approved by the
Promotor: Prof. dr. M.D. Kennedy and
Copromotor: Dr. B. Petrusevski

Composition of the Doctoral Committee:

Chairman	Rector Magnificus TU Delft
Vice-Chairman	Rector UNESCO-IHE
Prof. dr. M.D. Kennedy	UNESCO-IHE / TU Delft, promotor
Dr. B. Petrusevski	UNESCO-IHE, copromotor

Independent members:	
Prof. dr. G. J. Witkamp	TU Delft
Prof. dr. ir. W.G.J. van der Meer	Twente University
Prof. dr. M.S. Onyango	Tshwane University of Technology, South Africa
Prof. dr. M. Dimkic (Milan)	University of Novi Sad, Republic of Serbia
Prof. dr. M.E. McClain	TU Delft, reserve member

CRC Press/Balkema is an imprint of the Taylor & Francis Group, an informa business

Published by:
CRC Press/Balkema
Schipholweg 107 C, 2316 XC, Leiden, the Netherlands
Pub.NL@taylorandfrancis.com
www.crcpress.com – www.taylorandfrancis.com

ISBN 978-0-8153-9207-1.

Acknowledgement

I wish to thank ALMIGHTY GOD for the gift of life and for granting me HIS GRACE and opportunity to pursue this research work. I give ALMIGHTY GOD all the gratitude for the divine strength, abundant favour and for an incident free period of research. This research project was carried out at UNESCO-IHE Institute for Water Education with financial support from Dutch Government (DUPC, Program 1) project, for who I am greatly indebted to.

I wish to express my very special thanks and sincere gratitude to my promoters, copromoter and mentors (past and present): Gary Amy, Maria D. Kennedy, Branislav Pretrusesvki, Kebreab Ghebremichael and Richard Buamah for their patience and very valuable guidance in bringing this research project to a successful completion and the thesis in this current form. I am also grateful to Cyril Aubry of King Abdallah University of Science and Technology (KAUST) and Mr. Stefaan Heirman of TU Delft who helped with some aspects of characterization of adsorbent materials produced in this research. Many thank to Jan Schippers for some valuable advices at the initial stages of the laboratory work.

A number of Master of Science students greatly supported the laboratory work for the success of this research thesis, namely, L. Modestus, E. Mwapashi, Iddi Pazi, L.M Msenyele, Immaculata P. Misilma and Dava Augustus, and I do express my sincere gratitude to all, hoping that the effort will one day be helpful to fluoritic communities in developing countries. My gratitude also goes to Mr. Mufta Triban (intern from Libya) for his very hard work and great support to the laboratory work.

The completion of this thesis may not have been possible without the help of the UNESCO-IHE laboratory staff (past and present): Fred Kruis, Don van Galen, Frank Wiegman, Lyzette Robbemont, Ferdi Battes, Peter Heerings and Berend Lolkema. I am sincerely grateful for all their support, both in the laboratory and preparations for two field works in Ghana. Special thanks also go to other UNESCO-IHE staff, including

Jolanda Boots, Sylvia van Opdorp-Stijlen, Mariëlle van Erven, Anique Karsten and Felix Floor for all their tremendous assistance in academic and/or non-academic-related issues. I also say many thanks to Bianca Wassenaar, Bro. Alfred Larm and Paula Derkse who helped very much with formatting of the thesis and making it ready for printing. My special thanks to Jantinus Bruins who helped with the Dutch translation of the thesis summary.

I would like to thank my colleagues, whose company I enjoyed in the period of this work including, Jantinus Bruins, Valantine Umariwiya, Chol Abel, Mike Acheampong, Laurens Wellens, Mulele Nabuyanda, Nirajan Dhakai, Saaed Baghoth, George Lutterodt, Benjamin Botwe, Owusu Ansah and all the many Ghanaian students (past and present).

I thank the Community Water and Sanitation Agency (Ghana), staff of Water Research Institute (WRI) in Tamale (Ghana), the Bongo District Assembly (Ghana) and many individuals of Bongo Township, who assisted me during my field work for testing adsorbent material developed in this research for fluoride removal. Additionally, I thank the Church of Christ Rural Water Project and UNIHYDRO Ltd (Ghana) for their assistance during data collection and groundwater sampling in Northern region of Ghana.

Over the period of this work, I did worship at Mount Zion International Parish (MZIP) – Redeemed Christian Church of GOD (RCCG) in Delft and I wish to thank the leadership, his family and all the brethren for their warmness and prayers. I also thank the IHE Christain Fellow members (past and present) for their warmness and prayers.

Many thanks also go to Regina and Gerard Bilars, Alies vd Goot and Dick and, Mariette Bruijn for their kindness and warmness.

I thank my family and friends in Ghana for their support and prayers.

Finally, there may be more people who supported me directly or indirectly during this research and who are perhaps mistakenly not mentioned in this acknowledgements, I sincerely apologise to such persons. May GOD remember such persons and reward them for all the help and support they gave me.

This thesis is dedicated to the blessed memory of my beloved brother Ahmed Mohamed Salifu and also to all persons exposed to fluoride-related health hazards.

Abdulai Salifu

20 December 2016
Delft, The Netherlands

Summary

It was estimated as late as 2015 that, 663 million people worldwide still use unsafe drinking water sources, mostly in the least developed countries/regions including, sub-Saharan Africa and Southern Asia. A vast majority of the affected population are poor and live in rural areas. At the start of this research (2009), the population across the developing world without safe water sources was estimated at 884 million, and even though a lot was achieved by the end of the Millennium Development Goals (MDGs) deadline (2015), it is clear that a great deal still has to be done.

Access to safe drinking water is not only fundamental to human development and well-being, but is also recognized as a human right. The provision of safe potable water is considered critical and pivotal to the achievements of overall development, including adequate nutrition, education, gender equality and especially eradication of poverty in less developed countries.

Groundwater sources are generally known to be of good microbiological and chemical quality, and mostly require minimal or no prior treatment for use as safe drinking water sources. Hence its use for water supply is associated with low capital, as well as low operation and maintenance cost. It is therefore the most attractive source for drinking water supply in the often scattered, rural communities in developing countries. Problems can, however, occasionally arise with the chemistry of groundwater and render it unsafe, due to elevated concentrations of some elements that can have negative health impacts on the user. Provision of safe drinking water from groundwater sources in such situations therefore, requires some level of treatment. Fluoride is one of the water quality parameters of concern, the excess (beyond 1.5 mg/L, the World Health Organization (WHO) guideline value) of which contaminates groundwater resources in many parts of the world, and renders it not potable for human consumption, due to the related adverse health effects.

Even though an optimum concentration of fluoride (0.5 - 1mg/L) in drinking water is good for dental health and good bone development, the toxic effects on human health when consumed in excess amounts (beyond 1.5 mg/L), for long periods, have been known for centuries. The human health hazards of the consumption of excess fluoride include: the incidence of fluorosis, changes of DNA structure, lowering of IQ of children, interference with kidney functioning and even death when doses reach very high levels (about 150 - 250 mg/L).

Over 90 % of rural domestic water requirements in the Northern region of Ghana for instance (which was the area of focus of this study), is met from groundwater resources. Fluoride contamination of the groundwater in some parts of the region has, however, exposed the population in the fluoritic communities to fluoride-related health hazards. This has also resulted in the closure of otherwise many very useful drilled boreholes (wells) for water supply, in order to avoid the incidence of fluorosis and other related health effects. The closure of drilled (expensive) boreholes due to presence of excess fluoride do not only represent huge economic cost, but also hampers efforts of providing safe drinking water to the populace. As a consequence the population is forced to use unsafe surface water sources that are associated with the incidence of otherwise preventable, diseases such as cholera and diarrhea. Even though groundwater remains the most important source for rural water supply in the Northern region of Ghana, little is known about the factors (natural and/or anthropogenic) that control the groundwater chemistry and, hence the quality and source of fluoride contamination as well as its distribution.

Due to the permanent risk as well as the lack of known effective treatment for fluorosis and other related health hazards, defluoridation of fluoride-contaminated groundwater sources intended for drinking is a necessity, to avoid the ingestion of excess fluoride as a preventive measure. Several defuoridation technologies have been developed in many places around the world, some of which are described as "Best Available Technologies" (BATs). The current methods, however, mostly have some limitations which generally make their use unsustainable and/or unacceptable under most conditions, particularly in remote areas in developing countries. This include for instance: (i) the Nalgonda technique, which is popular in some Asian countries but is known to have limited efficiency (up to about 70 %), requires careful dosing of chemicals and close monitoring to ensure effective fluoride removal, hence demanding labour, skills and time that are usually problematic under rural conditions in developing countries; (ii) the contact precipitation process, which is still under study, and moreover the reaction mechanism for the defluoriidation process is thought only to be feasible with use of bone charcoal as a catalyst. Bone charcoal is however not culturally acceptable in some societies due to local taboos and beliefs; (iii) adsorption using activated alumina as adsorbent media, which is known to be expensive especially for developing countries, (iv) adsorption with bone charcoal as adsorbent media, which is not acceptable in many places as earlier mentioned, and, (v) reverse osmosis (RO), which has high capital and operational cost, require specialized equipment, skilled labour and a continuous supply of energy.

Due to the negative health impacts of excess fluoride in drinking water, however, the search for an appropriate technology for its removal from contaminated-groundwater still remains very critical. Among the available fluoride removal techniques, the adsorption process is generally considered as one of the most appropriate, particularly for small community water source defluoridaton. This is due to its many advantages including flexibility and simplicity of design, relative ease of operation, and cost- effectiveness as well as its applicability and efficiency for contaminant removal even at low concentrations. The appropriateness of the adsorption technology, however, largely depends on availability of a suitable adsorbent. Several adsorbent materials have been developed and tested,   mostly in the laboratory, for the treatment of fluoride-contaminated water including: manganese-oxide coated alumina, bone charcoal, fired clay chips, fly ash, calcite, sodium exchanged montimorillonite-Na^+, ceramic adsorbent, laterite, unmodified pumice, bauxite, zeolites, fluorspar, iron-oxide coated sand, calcite, activated quartz and activated carbon. While some of these adsorbent materials have shown certain degrees of fluoride adsorption capacity, the applicability of most is limited either due to: lack of socio-cultural acceptance, non regenereable nature, and therefore may not be cost-effective, high cost and/or effectiveness only under extreme pH conditions. This may require pH adjustment and consequently additional capital, operation and maintenance cost, and could limit feasibility of such a fluoride removal technology in remote rural areas of developing countries. Some of the studied defluoridation materials are also available in the form of fine particles or powders, with the potential of clogging and/or low hydraulic conductivities when applied in fixed bed adsorption systems. The search for appropriate alternative fluoride adsorbents therefore remains of interest.

The overall goal of the study was therefore twofold: (i) to study the groundwater chemistry in the Northern region of Ghana with focus on the occurrence, genesis and distribution of fluoride-contaminated waters in the eastern corridor of the region (which is the most fluoritic part), and (ii) to contribute to the search for an appropriate and sustainable fluoride removal technology for the treatment of fluoride-contaminated groundwater for drinking water production in developing countries.

In order to accomplish the first component of the research goal, the climate, geology, mineralogy and hydrogeology of the study area was reviewed in a desk study. Three hundred and fifty seven (357) groundwater samples taken from boreholes drilled in the study area, were analyzed for the chemical data using standard methods. Univariate statistical analysis, Pearson's correlation and principal component analysis (PCA) of the chemical data, using the SPSS statistical software package, as well as Piper graphical classification using the GW chart software, and thermodynamic calculations with PHREEQC, were used as complementary approaches to gain an insight into the groundwater chemical composition, and to understand the dominant mechanisms influencing the occurrence of high-fluoride waters in the area. The geo-referenced groundwater chemical data were further analyzed using ArcGIS software to determine the spatial distribution of fluoride in groundwater at the sampled points of the study area. Inverse distance weighting interpolation (IDW) (using ArcGIS), was also used to examine the fluoride distribution in the study area and to help predict the fluoride levels at non-sampled points as well.

The fluoride concentrations in the 357 groundwater samples taken from the area were found to range between 0.0 and 11.6 mg/L, with a mean value of 1.13 mg/L. A relatively high percentage (23%) of the samples were found to have fluoride concentrations exceeding 1.5 mg/L, the WHO guideline value for drinking water. Based on the piper graphical classification, six groundwater types were identified in the study area: $Ca-Mg-HCO_3$, $Ca-Mg-SO_4$, $Na-Cl$, $Na-SO_4$, $Na-HCO_3$ and mixed water type. PCA performed on the groundwater chemical data resulted in 4 principal components (PCs) explaining 72% of the data variance. The PCs, which represented (or explained) the predominant geochemical processes controlling the groundwater chemistry in the study area, showed that the processes include: mineral dissolution reactions, ion exchange and evapotranspiration. PHREEQC calculations for saturation indices of the groundwater samples revealed that they were largely saturated with respect to calcite and under-saturated with respect to fluorite, suggesting that dissolution of fluorite may be occurring in the areas where the mineral is present. A review of the principal component analysis (PCA) results, coupled with an evaluation of the equilibrium state of the groundwater based on the calculated saturation indices, suggested that the processes controlling the overall groundwater chemistry in the area also influenced the fluoride enrichment. The fluoride-rich groundwaters in the study area were found to occur predominantly in the Saboba and Cheriponi districts, and also in the Yendi, Nanumba North and South districts of the Northern region of Ghana. These areas are underlain by the Obossom and Oti beds, comprising mainly of sandstone, limestone, conglomerate, shale, arkose and mudstone. Results of the conducted hydrochemical analysis showed that apart from the boreholes with high concentrations of fluoride (beyond 1.5 mg/L), groundwater in the study area, based on the limited parameters analyzed is chemically acceptable and suitable for domestic use.

GIS analysis of the geo-referenced groundwater data resulted in a map showing the spatial distribution of fluoride in groundwater at the sampled points of the study area, and also a prediction map that may help determine the fluoride levels at the non-sampled parts of the area. The information provided by the study may help in strategic planning for the provision of improved water sources to the populace in the study area, in either case of high fluoride groundwater, or groundwater wells with insufficient fluoride (for prevention of dental caries).

The second component of the research goal was accomplished through a combination of studies related to: a laboratory synthesis of alternative fluoride adsorbent materials, an assessment of their fluoride adsorption performance (i.e. kinetics and equilibrium capacity), a study of the mechanisms of fluoride removal and, a comparison of the performance to both activated alumina (AA), the industrial fluoride adsorbent, and also to other developed fluoride adsorbents reported in literature. The study also included the possibility of regenerating the synthesized materials when exhausted for its economic and practical viability, safe disposal of the spent adsorbent into the environment (when it can no longer be used), and a field screening of the capability of the synthesized adsorbent to treat natural fluoride-contaminated groundwater.

The synthesis of a fluoride adsorbent was accomplished by exploring the possibility of modifying the physico-chemical properties of the surfaces of locally available materials, that contributes to a low cost of adsorbent production and sustainability, when in use in developing countries. Three materials, i.e. pumice, bauxite and wood charcoal, which are indigenous and readily available in many developing countries, were used as base material for the surface modification process. The use of different indigenous materials for the surface modification process was aimed at investigating the influence of the starting base materials on fluoride removal performance of the modified version, and, hence at determining which is more suitable as a precursor material. Surface modification of the indigenous materials was accomplished by an Al coating/Aluminol (AlOH) functionalization process, using 0.5 M $Al_2(SO_4)_3$ as the Al – bearing solution, and, exploring the hard soft acid base (HSAB) concept. Due to its characteristics, Al (III) is classified as a hard acid, while F- is categorized as a hard base. Al (III) therefore has good affinity for fluoride in accordance to the HSAB concept, hence its choice for the modification of the indigenous materials surfaces, as alternative materials for fluoride adsorption. Three different types of Al modified adsorbents were produced, i.e. Al oxide coated pumice (AOCP), granular Al coated bauxite (GACB), and Aluminol functionalized wood charcoal (AFWC). For bauxite, the influence of the synthesis process conditions on the effectiveness of the coating process, and the consequent effect on the fluoride adsorption efficiency of the produced adsorbent (i.e GACB), was investigated, which was aimed at optimizing the synthesis process. These conditions included different coating pH, and different process temperatures for thermal pre-treatment of bauxite prior to the aluminum coating. For wood charcoal, the effect of the source of wood charcoal (i.e. from 4 sources) on the fluoride removal performance of the produced adsorbent was also investigated. Several characterization techniques were employed for studying the physico-chemical properties of the synthesized fluoride adsorbents. These included: X-ray power diffraction (XRD), X-ray fluorescence (XRF) analysis, scanning electron microscopy (SEM), Fourier transform infrared (FTIR) analysis, Raman spectroscopy, energy-dispersive X-ray (EDX) analysis, BET specific surface area analysis, mass titration and potentiometric titration. This was helpful in informing and guiding the fluoride adsorbent material engineering process. Regeneration of the produced adsorbents, when exhausted, was accomplished by creating new Al-based active sites on the surfaces of the fluoride-saturated adsorbent through an Al re-coating/aluminol re-functionalization process.

Batch fluoride adsorption and/or laboratory-scale continuous flow column experiments were conducted to either evaluate the fluoride removal efficiencies of the produced adsorbents, or to determine their kinetic properties and/or fluoride adsorption capacities. Several kinetic models were used for the interpretation of the kinetic data; which included the pseudo-first-order and pseudo–second-order kinetic models, Banghams equation, Elovich and the Weber and Morris intra-particle diffusion models. Different models give different information which can complement each other for a better understanding of the properties of the synthesized adsorbent as well as the nature of the adsorption process. Similarly, interpretation of the equilibrium data was accomplished using several isotherm models including; Langmuir, Freundlich, Temkin, Elovich, BET, Generalized, Dubinin-Raduskevich and Redilich-Perterson equations. The continues flow column experimental data in the form of breakthrough curves were also analyzed/modelled using three well known models: Thomas, Adams-Bohart and the bed-depth service models. Due to the inherent errors associated with linearization, both linear regression and non-linear optimization techniques (with error analysis), were employed to determine best-fit and the related model parameters. Due to the complex nature of adsorption processes, the kinetic and equilibrium modelling in combination with the Raman and FTIR spectroscopic analysis as well pH_{pzc} measurements and thermodynamic calculations were used as complementary approaches to gain insight into the mechanisms involved in fluoride removal onto the produced materials. The effects of pH and/or co-ions (sulfate, chloride, bicarbonate, nitrate and phosphate) as well as long-term storage of adsorbent on the performance of the three produced fluoride adsorbent materials, were examined in batch adsorption experiments. The effect of particle size on the performance of the Al modified wood charcoal was studied under continuous flow conditions.

For all indigenous materials (pumice, bauxite and wood charcoal), a modification of their particle surfaces by the Al coating/aluminol (AlOH) functionalization process was found effective in creating Al-based surface active sites for fluoride adsorption from aqueous solutions, in accordance to the hard and soft acid base (HSAB) concept. GACB, AOCP and AFWC were all able to reduce fluoride concentration in model water from 5.0 ± 0.2 to ≤ 1.5 mg/L (WHO guideline) within 32, 1 and 0.5 h, respectively, indicating that aluminol functionalized wood charcoal (AFWC) possessed superior kinetics of fluoride removal and was also more efficient. The thermal pre-treatment of bauxite prior to the aluminum coating contributed significantly to an increase of the textural properties (i.e. surface area and pore volume), the effectiveness of the coating process, and hence the fluoride removal efficiency of the produced GACB, compared to Al coating of the untreated bauxite. The optimum synthesis conditions were established at coating pH of 2, with thermal pre-treatment at 500°C.

The source of wood charcoal was also found to influence the fluoride removal capability of the corresponding produced AFWC. This therefore enabled the selection of the most suitable wood charcoal as precursor material for the aluminol functionalization process for further work.

Kinetic and equilibrium fluoride adsorption experimental data for all the synthesized adsorbents conformed reasonably well to the kinetic and isotherm models used for the modelling and interpretation of the kinetic and equilibrium data. The related kinetic and isotherm model parameters were estimated, and are useful for design purposes. The kinetic and isotherm analysis, pH_{pzc} measurements and thermodynamic calculation, as well as the FTIR and Raman analyses, complementing each other suggested that the mechanism of fluoride removal is complex, and involves both physical adsorption and chemisorption processes.

Based on results from laboratory kinetic adsorption experiments, it was observed that at a neutral pH of 7.0 ± 0.2, which is a more suitable condition for groundwater treatment, fluoride adsorption by AOCP was fairly faster than that of the industrial standard fluoride adsorbent, activated alumina (AA), with similar particle size range (i.e. $0.8 - 1.12$ mm).

Moreover, the fluoride adsorption capacities of the Al modified adsorbents, based on Langmuir isotherm estimates from the batch equilibrium adsorption data, were found to be either comparable or higher than that of some reported fluoride adsorbent (for adsorption at pH 7.0 ± 0.2). The three developed adsorbents were therefore considered to be promising and could be used for the treatment of fluoride-contaminated groundwater. The applied procedures in this study could therefore be a useful approach for synthesizing effective adsorbents for use in fluoritic areas of developing countries, with a possibility of reduction in production cost, especially where the indigenous base materials (pumice, bauxite, wood charcoal) are locally available. Use of locally available base materials will also contribute to sustainability.

Both AOCP and AFWC exhibited good fluoride adsorption efficiencies within the pH range 6 to 9, which makes it possible to avoid pH adjustment with the associated cost and operational difficulties, especially if these adsorbents are to be used in remote areas of developing countries. The presence of nitrate, bicarbonate, chloride and phosphate, at concentrations commonly found in groundwater, showed either no or only negligible effects on the fluoride removal by AFWC and GACB. Sulfate, however, showed some retardation of the fluoride removal by AFWC and GACB, even though the effect was milder on the performance of AFWC.

In a further assessment of the effectiveness of AOCP in laboratory-scale column experiments, which is a more useful approach for obtaining parameters for optimizing the design of full scale treatment systems, the adsorbent was similarly capable of reducing fluoride concentration of model water from 5.0 ± 0.2 mg/L to ≤1.5 mg/L, under the continues flow conditions. AOCP in particle size range 0.8 – 1.12 mm, treated 165 bed volumes (BV) of model water, at the time of breakthrough. The fluoride-AOCP experimental breakthrough data could be well described by the Thomas, Adams-Bohart and bed-depth service time (BDST) models. The derived model parameters are useful for up-scaling purposes for the design of AOCP defluoridation filters, without need for further experimentation. Regeneration of exhausted AOCP was found feasible. The adsorption capacity of AOCP after the 1st regeneration cycle was not only fully (100 %) restored, but was observed to *increase* by more than 30%, indicating the effectiveness/usefulness of the regeneration procedure. The regenerability of fluoride-saturated AOCP contributes to its economic and practical viability. In a similar assessment of the performance of AFWC (particle size range 0.8 – 1.12 mm) in laboratory-scale column experiments, the adsorbent treated 219 bed volumes of water before breakthrough, which indicated AFWC has a higher fluoride adsorption capacity than AOCP (an increase of approximately 30 %), under the continuous flow conditions. AFWC was therefore used for further work. A reduction of the particle size range of AFWC to 0.425 - 0.8 mm, aimed at further optimizing the performance, resulted in the treatment of 282 bed volumes of water before breakthrough, which represented a further increase of 28%.

Granular aluminum coated bauxite (GACB) was, however, found to reduce in its fluoride removal efficiency after 8 months of storage whereas the performances of AFWC and AOCP remained the same after 8 and 12 months under similar storage conditions, respectively. Moreover to enhance the performance of GACB, a thermal pre-treatment of the bauxite base material was required prior to the Al coating, which will require an energy source and the use of special calcination equipment, all of which will increase the production cost. The performance of GACB was therefore not assessed further under laboratory-scale continues flow conditions in this study.

Since AFWC showed better performance than AOCP under similar laboratory continuous flow conditions, its efficacy and performance in a real world situation was further tested in a field pilot study in Bongo town, which is within the fluoritc areas in Ghana. AFWC (particle size range 0.8 – 1.12 mm, which was available at the time of the field test) was capable of reducing a fluoride concentration of 4.88 mg/L in natural groundwater under field conditions to ≤ 1.5 mg/L. Similar number of bed volumes of treated water untill breakthrough as obtained under laboratory conditions, was achieved in the field (i.e 208 BVs). The laboratory fluoride removal was thus reproducible under field conditions. AFWC was also found to be regenerable, when exhausted. In a similar field screening of the performance of regenerated AFWC (RAFWC), the fluoride adsorption capacity was found to increase by about 40 % compared to that of the freshly produced AFWC (i.e. 295 bed volumes treated before breakthrough). The trend of improved removal performance after regeneration was thus similar to that observed for AOCP and RAOCP.

A comparison of the fluoride removal performance of AFWC with a particle size range 0.8 – 1.12 mm, with that of activated alumina (AA), the industrial fluoride adsorbent, under similar field conditions indicated a higher performance of the AA. The performance of AA was, however, found to be depended on the type or grade employed. The AA grade previously compared to AOCP (under laboratory conditions) was of similar particle size range of 0.8 – 1.12 mm, and was supposed to be regenerable (according to the manufacturer). The particle size range of the AA grade used in the field study was, however, much finer (i.e. 0.21 – 0.63 mm) than 0.8 -1.12mm, which presumably contributed to its higher performance. Moreover, the AA grade tested in the field was (according to the supplier's e-mail communication) effective but not regenerable, and has to be used once and disposed of, when exhausted. Based on literature information, however, the application of AA for water defluoridation in developing countries can only be cost-effective if it can be regenerated and applied multiple cycles.

Furthermore, characterization of fluoride-saturated (spent) AFWC using the US-EPA toxicity characterization procedure (TCLP) indicated that it is non-hazardous, and could be disposed of in simple landfill, whereas, spent AA may require further handling before disposal in order to avoid environmental contamination, which could increase operational costs.

Even though the current fluoride adsorption capacity of AFWC still requires further improvement, the field performance of AFWC and RAFWC is encouraging, and it is too early to conclude superior performance of AA over AFWC1`, since the AA screened in the field cannot be regenerated, while AFWC is regenerable. Moreover, AFWC shows potential for regeneration with increasing performance after each regeneration cycle. AFWC could therefore possibly be developed further and it can most likely contribute to the provision of safe drinking water to some of the 663 million people still using unimproved sources, especially those living in rural fluoritic areas of developing countries.

Table of Contents

1

General introduction

1.1   Background

Significant progress was made during the Millennium Development Goal (MDG) period (1990 – 2015), towards the achievement of the global target for safe drinking water, which was met in 2010, much ahead of the MDG deadline of 2015 (UNICEF and WHO, 2015). In spite of the good achievement and, with 91 % of the global population now having access to improve drinking water, it was also observed that some developing/least developed regions including Southern and Central Asia, North Africa, Oceania and sub-Saharan Africa were unable to meet the drinking water target. Moreover, huge disparities with regards to the global access to safe drinking water were also observed, including inequalities such as the gap between the urban population (who are better served) and the rural population, the gender burden of water collection and the gap between the richest and the poorest and most venerable segments of society, who lack access to improved water services. It was observed that 8 out of 10 people (i.e 80 %) who still lack access to safe drinking sources are rural dwellers (UNICEF and WHO, 2015).

As early as 2015, it was estimated that 663 million people worldwide still use unsafe drinking water sources, mostly in the least developed countries/regions including sub-Saharan Africa and Southern Asia, a vast majority of who are mostly poor and also live in rural areas.

At the start of this study (2009), the population across the developing world without safe water sources was estimated at 884 million (UNICEF, 2009; MacDonalds, 2009), and even though a lot was achieved by the end of the MDGs deadline (2015), it is clear that a great deal still require to be done.

Access to safe drinking water is not only fundamental to human development and well-being, but is also recognized as a human right (UN General Assembly, 2010; UNICEF & WHO, 2015). The provision of safe potable water is considered critical and pivotal to the achievements of overall development, including adequate nutrition, education, gender equality and especially eradication of poverty in less developed countries (Pollard et al., 1994; UNICEF and WHO, 2015).

One of the main problem concerning the provision of safe drinking water in developing countries is often associated with the poor quality of the water source and the need for treatment. There are many problems associated with water treatment in developing countries. These include:

- high investment as well as operation and maintenance cost,

- complexity of some processes that require the use of special equipment, electrical energy and skilled personnel which are mostly not available in rural areas (Dysart, 2008).

- environmental concerns with regards to disposal generated waste (Boddu et al., 2008).

Due to economic constraints, the development of low-cost and appropriate water treatment technologies is deemed very necessary (Pollard et al., 1994).

1.2 Groundwater, fluoride contamination, the benefits and pathophysiology

Groundwater sources are generally known to be of good microbiological and chemical quality and mostly require minimal or no prior treatment for use as safe drinking water sources. Hence its use for water supply is associated with low capital as well as low operation and maintenance cost. It is therefore the most attractive source for drinking water supply in the often scattered rural communities in developing countries (Buamah et al., 2008; MacDonald and Davies, 2000; Gyau-Boakye and Dapaah-Siakwan, 1999). Problems can, however, occasionally arise with the chemistry of groundwater, due to elevated concentrations of some elements, which can have negative health impacts on the user (MacDonald, 2009). Provision of safe drinking water from groundwater in such situations therefore require some level of treatment. Fluoride is one of the water quality parameters of concern that contaminates groundwater resources in many parts of the world and renders it not potable for human consumption due to adverse health effects.

Fluoride is known to have both beneficial and detrimental effects on health, depending on the dose and duration of exposure (Mjengera and Mkongo, 2009; WHO, 2011; Madhnure et al., 2007; Ma et al., 2007; Biswsa et al. 2007; Fawell et al. 2006; Nagendra, 2003). For instance the unique ability of the chemical to inhibit, and even reverse negative health effects with regards to the tooth has been well observed (Whitford, 1996). Ingestion of optimum concentrations of fluoride (about 0.5 – 1.5 mg/L) in drinking water can prevent the incidence of dental caries, particularly in children up to age 8. It prevents the tooth decay by inhibiting the production of acid by decay-causing bacteria. These orally present bacteria (most prominently, *streptococcus mutans* and *streptococcus sobrinus*, and *lactobacilli*), consume food debris or sugar (sucrose) on the tooth surface (from the food we eat) for their own source of energy, and in the process convert them to lactic acid through a glycolytic process known as fermentation. These organisms are capable of producing high levels of lactic acid and when in contact with the tooth, can cause the dissolution/breakdown of minerals from the enamel (a highly mineralized cellular tissue), which plays a very important role in protecting the teeth from decay. When the tooth enamel loses its mineral content (i.e demineralization of mostly hydroxyapatite and calcium phosphate), it becomes weak and vulnerable to decay. Thus inhibition of the action of the decay-causing bacteria (by fluoride) from creating the required acidic environment around the enamel, beneficially helps to prevent the chemical processes (i.e mineral dissolution/breakdown) leading to tooth decay. Moreover, fluoride is also known to be a re-mineralization agent which can enhance the replacement of lost minerals from enamels that has been attacked, and thus reverse the formation of dental caries. Fluoride can bind to hydroxyapatite crystals in the tooth enamel and the incorporated fluorine makes the enamel stronger and more resistant to demineralization, hence resistance to decay (Whirtford, 1996).

Intake of excess fluoride (beyond 1.5 mg/L) for long periods can, however, result in negative human health effects. Fluoride has several mechanism of toxicity (Firempong et al., 2013; Shin, 2016; Whirtford, 1996). When it enters into the human body, mainly through the intake of water and to some extent food and dental products, about 75 – 90 % is adsorbed (Harder, 2008; Shomer, 2004; Fawell et al., 2006). Ingested fluoride ions initially acts on the gastrointestinal musoca to form hydrofluoric acid (HF) in the stomach by combining with hydrogen ions under the acidic condition in the stomach. The formation of hydrofluoric acid leads to nausea, diarrhoea, vomiting, gastricintestinal irritation and abnominal pains. About 40 % of the ingested fluoride is adsorbed from the stomach as HF. Fluoride not adsorbed in the stomach is adsorbed in the instestine. Once absorbed into the blood stream, fluoride readily distributes throughout the body and tend to accumulate in calcium rich areas such as bone and teeth (dentin and enamel) (Fawell et al., 2006; Firempong et al., 2013; Shin, 2016; Gessner et al., 1994). At moderately high levels (1.5 – 4 mg/L) of ingestion, it leads to dental fluorosis, particularly in children. According to Whirtford (1996), even though the mechanisms underlying the development of dental fluorosis are not well understood, there is evidence that the processes probably involve effects on the ameloblasts, which deposit tooth enamel. Ameloblast are cells present during tooth development (in childhood), and secretes the anamel proteins (i.e enamelin and amelogenin), that mineralizes to form the tooth enamel. These cells are observed to be very sensitive to their environment, and bodily stressors (during childhood) can affect their function hence, cause interruption in enamel production. Presumably exposure of children (between the ages of 2 to 8 years old) who are still undergoing mineralization in the permanent teeth to excess fluoride (1.5 – 4 mg/L), is a type of stressor that disrupts the enamel production and results in the development of dental fluorosis (Firempong et al., 2013; Whirtford, 1996; Fawell et al., 2006). Dental fluorosis, which is characterized by discoloured, blackened, mottled or chalky-white teeth, is by far the most common manifestation of chronic use of high-fluoride water. A person affected by dental fluorosis is an indication of overexposure to fluoride during childhood when the teeth were developing (Fawell et al. 2006). These effects are, however, not apparent if the teeth are already fully grown prior to the fluoride overexposure. Therefore if an adult shows no signs of dental fluorosis, it does not necessarily mean his or her fluoride intake is within safety limits and could be at risk of other fluoride-related health hazards.

At higher levels (> 4 mg/L), fluoride may disrupt the mineralization of bones leading to severe and permanent bone and joint deformations, commonly named skeletal fluorosis (Fawell et al., 2006; Firempong et al., 2013; Shin, 2016). Early symptoms of skeletal fluorosis include sporadic pain and stiffness of joints. Headache, stomach-ache and muscle weakness can also be warning signs of skeletal fluorosis. The next stage is osteosclerosis (hardening and calcifying of the bones) and finally the spine, major joints, muscles and nervous system become damaged (UNICEF, 2009b).

Ingestion of excess fluoride can also cause other ailments or adverse effects besides skeletal and dental fluorosis, i.e., non-skeletal menifesations. These are, however, normally overlooked because of the misconception that excess fluoride does only affect bone and teeth (Nagendra, 2003; Mjengera and Mkongo, 2009).

According to the following workers/authors: Shin (2016), Firempong et al. (2013), Valdez-Jimenez et al. (2010), Fawell et al. (2006), Meenakshi et al. (2006), Yu et al., 2008, Nagendra (2003) and Gessner et al. (1994), other adverse effects of ingesting high levels of fluoride include:

- abortion or still birth, due to exposure of pregnant mothers to chronic fluoride poisoning,
- infant mortality due to calcification of blood vessels during breast feeding (from mothers who consume high concentrations of fluoride),
- effects on brain tissues similar to Alzheimer's disease,
- neurological manifestation (nervousness),
- very painful skin rashes,
- depression,
- male sterility,
- low hemoglobin level,
- growth retardation of children,
- alteration in the functional mechanism/multiple failure of human organs (e.g. liver, kidney, etc), due to severe fluoride toxicity,
- the element can accumulate in calcium-rich areas of the body where it binds with calcium, resulting in hypocalcaemia i.e. low calcium levels in the blood serum, which may cause a number of health hazards including numbness, seizures and cardiac arrest.
- direct cytotoxic effects (i.e. being toxic to cells) and, hence interference with a number of enzyme systems, including disruption of the following:
  - oxidative phosphorylation, which is the metabolic pathway by which enzymes are used by cells to oxidize nutrients to release energy that is used to reform

adenosine triphosphate (ATP) energy. The ATP energy is required/essential for many living processes, such as muscle contraction and nerve impulses,

➢ neurotransmission (i.e. the process of communication between brain cells), which can alter the way the effected person think, feel and behave.

Several studies also suggest exposure to fluoride may adversely affect the intelligence quotient (IQ) of children (Valdez-Jimenez et al., 2010; Rocha-Amador el al., 2007; Shomar et al., 2004; Lu et al., 2000; Zhao et al., 1996). For instance, it was found from an evaluation conducted in China that the average IQ (= 97.69) of children from a known fluoritic village with a fluoride concentration of 4.12 mg/L in the water supply, was significantly lower than the average IQ (= 105.21) of children living in a non-endemic village, with a fluoride concentration of 0.91 mg/L in the water supply system. The risk of impaired development of a child's intelligence is observed to be due to exposure of an embryo to high levels of fluoride, arising from exposure of pregnant/potential mothers to chronic fluoride poisoning. Fluoride in the maternal blood in such cases can pass through the placenta to the foetus, and the element is reported as being able to penetrate the blood-brain barrier of the foetus and accumulate in the cerebral tissues, where it can cause detrimental biochemical and functional changes in the developing foetus/human brain. This can consequently result in lowered IQ and other child abnormalities/complications (Valdez-Jimenez et al., 2010; Rocha-Amador el al., 2007; Shomar et al., 2004; Lu et al., 2000; Zhao et al., 1996).

Accumulation of high levels of fluoride in the body over long periods of time has also been associated with changes in DNA structure in some cases (Maliyekkal et al., 2006).

Death may occur, usually from either respiratory paralysis or dysrhythmia (i.e. changes in the regular beat of the heart) or cardiac arrest, when the fluoride dose reach very high levels (Table 1.1) (Firempong et al., 2013; Shomar et al., 2004; Gessner et al., 1994). For instance, a typical case of human fatality in an Alaskan village in the United States of America (USA) due to acute fluoride poisoning was reported in 1994, where the concentration of the ingested fluoride was believed to be 150 mg F/L (Gessner et al., 1994).

Table 1.1 Indication of fluoride concentrations in drinking water and related health effects

Fluoride conc. (mg/L)	< 0.5	0.5 -1.5	1.5 - 4	>4	>10	>50	>100	>120	About 50 - 250
Health effects	Dental caries	Promotes dental health/ good bone development	Dental fluorosis	Dental/ skeletal fluorosis	Crippling fluoriosis	Thyroid changes	Growth retardation	Effects on kidney	Death

Beside the health effects, exposure to drinking water with high fluoride may also have serious psychological and social consequences including matrimonial problems of young adults (Shomar et al. 2004; Nagendra, 2003). For instance it has been observed that, younger people in some endemic communities in the Northern region of Ghana, showing symptoms of high dental fluorosis (severe pitting and coloration of teeth) find it difficult to smile comfortably in public. They also have difficulty finding spouses outside of their communities.

At a given a concentration, the effects of fluoride in drinking water is higher in places with high temperatures due to higher consumption of water, hence more ingestion of excess fluoride (Mjengera and Mkongo, 2009). Figs 1.1 & 1.2 show severe cases of skeletal fluorosis from India. Fig. 1.3 also shows the incidence of fluorosis in Ghana.

Figure 1.1: Severe skeletal fluorosis in India. India (Source: UNICEF, 2004)

Figure 1.2: Severe skeletal fluorosis in (This lady is only 40 years old, but looks 80 years!!) (Source: UNICEF, 2001).

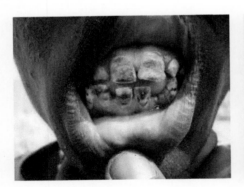

Figure 1.3: Incidence of dental fluorosis (left) and symptoms of skeletal fluorosis (right) in the Saboba/Cheriponi district of the Northern region of Ghana (Source: CWSA-NR, 2007).

1.3   High fluoritic regions

Fluoride contamination of groundwater and hence exposure to the risk of incidence of fluorosis and other related health hazards is a worldwide problem, including various countries in Africa, Asia, Europe as well as the USA and Australia (Maliyekkal et al., 2006; Fawell et al., 2006). Conservatively, it is estimated that, tens of millions of people in at least 25 nations around the world are affected by fluorosis (UNICEF, 1999b). China and India are among the most affected (Maliyekkal et al., 2006) and other countries such as Ethiopia, Kenya, Ghana and Tanzania have serious problem related to fluoride contamination (Fawell et al., 2006).

Areas where high fluoride waters have been found to occur around the world include large and extensive geographical belts associated with the following:

(i) Sediments of marine origin in mountainous areas; examples of such areas include Iraq, Iran, the Mediterranean region, the southern parts of former Soviet Union, southern parts of USA and southern Europe (Fawell et al. 2006),

(ii) Areas with volcanic activity as found in the East African Rift valley system. Many of the lakes in the Rift Valley system, particularly the soda lakes are reported as having exceptionally very high fluoride concentrations (over 2000 mg/L) (Walther, 2009; Fawell et al., 2006; Hurtado et al., 2000; Apambire et al., 1997), and

(iii) Areas with granitic and gneissic rocks as found in places such as India, Pakistan, China, South Africa and West Africa (Fawell et al., 2006; Apambire et al., 1997).

Fig 1.4 shows a probability map of occurrence of high fluoride (>1.5 mg/L) globally, hence probability of exposure of the populace to incidence of fluorosis and other related health hazards due to intake of excess fluoride in drinking water.

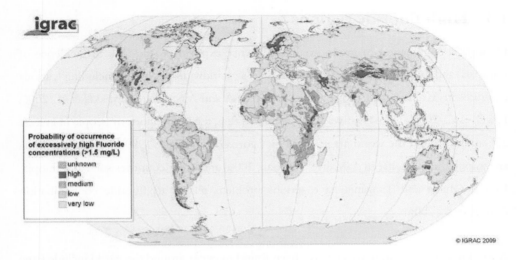

Fig 1.4: Probability map of occurrence of high fluoride (>1.5 mg/L) (Source: IGRAC, 2009).

1.4   Available fluoride removal technologies

Because of the permanent risk and also the lack of known effective treatment for fluorosis, fluoride removal from contaminated drinking water is a necessity, to avoid ingestion of excess fluoride as a preventive measure (Sahli et al., 2007; Sarkar et al., 2007).

Several defuoridation technologies have been developed in many places around the world, some of which are described as "Best Available Technologies" (BATs). The current methods, however, have some limitations which generally make their use unsustainable under most given conditions, particularly in remote areas in developing countries (Ayamsgna et al., 2008; Feenstra et al., 2007; Fawell et al., 2006; Meenakshi et al., 2006; Dahi, 1999; Attanayake et al., 1995).

The common fluoride removal techniques include those based on sorption process, coagulation-flocculation-filtration, contact precipitation and membrane filtration processes (Dysart, 2008; Meenakshi et al., 2006; Fawell et al., 2006).

1.4.1 Techniques based on sorption process

The most popular fluoride removal technologies based on the sorption process include fluoride adsorption on activated alumina and bone charcoal.

1.4.1.1 Activated alumina (AA)

The technology which uses activated alumina as a filter media, has been described by the United States Environmental Protection Agency (US EPA) as a Best Available Technique for water defluoridation (Dysart, 2008). The use of the technique which started in the 1930's, remains popular and has become the method of choice for water defluoridation in industrialized countries including the USA and Australia. The technology has also gained attention in some less developed countries particularly in India where its use is propagated in villages with the support of UNICEF. The technique has a fluoride removal performance of about 85-95% (Dysart, 2008; Feenstra et al., 2007; Meenakshi et al., 2006; Iyengar, 2003).

The activated alumina (Fig. 1.4) is aluminum oxide (Al_2O_3) grains prepared to have a large sorptive surface. The media is prepared by a controlled thermal treatment of granules of hydrated alumina to produce a highly porous media. It consists essentially of a mixture of amorphous and crystalline phases of aluminum oxide referred to as aluminum trihydrate. Alumina has a high pH point of zero charge ($pH_{PZC} \sim 8.2$) indicating it has an adsorption affinity for many negatively charged particles including fluoride ions (Dysart, 2008; Fawell et al., 2006).

Fig. 1.5 Activated alumina grains

The mechanisms of fluoride removal by the activated alumina method has been described as complex, mainly involving an adsorption and an ion exchange processes where hydroxyl ions (OH$^-$) are exchanged for fluoride ions in solutions (Dysart, 2008; Feenstra et al., 2007; Fawell at al., 2006).

The fluoride removal capacity of alumina is highly sensitive to pH, hence defluoridation systems based on activated alumina often need pH adjustment in their operations in order to optimize the fluoride removal (Dysart, 2008; Fawell at al., 2006; Meenakshi et al., 2006). Generally the optimum fluoride removal has been found to occur in the range of pH 5 to 6 (Dysart, 2008; Meenakshi et al., 2006; Ku and Chiou, 2002). At pH above 7 the uptake of fluoride by the activated alumina media decreases. This is because the exchange reactions between surface hydroxyl groups and the adsorbing fluoride ion become less favorable, as silicates and hydroxides in solution become a stronger competitor of the fluoride ion for exchange sites on the activated alumina surface. Increasing pH also favors the electrostatic repulsion between the negatively charged alumina surface and the anionic fluoride. At pH< 5, activated alumina gets dissolved in the acidic environment leading to loss of the media and hence a decrease in the uptake of fluoride (Dysart, 2008; Meenakshi et al., 2006; Ku and Chiou, 2002).

When activated alumina is used for the fluoride removal it eventually becomes exhausted and has to be regenerated. The regeneration process is carried out by exposing the media to an alkaline solution, typically caustic soda (NaOH), to strip of the fluoride and restore the removal capacity. As the fluoride removal capacity is strongly dependent on pH, an acid solution typically sulfuric acid or hydrochloric acid is subsequently used to neutralize and reactivate the media. During each regeneration cycle about 5-10% of the alumina is lost, and usually results in a reduction of the capacity of the media by 30-40%. The media often has to be replaced after 3-4 regenerations (Dysart, 2008; Fawell at al., 2006).

The activated alumina technique has been noted to have the following limitations/shortcomings:

- The technology is expensive to acquire, operate and maintain for sustainable use in developing countries (Fawell at al., 2006; Ayamsgna et al., 2008);

- Because the process works best in a narrow pH range, chemical feed equipment, storage and handling of corrosive chemicals, and skills are required for pH adjustment if the fluoride removal process is to be optimized. These additional requirements, have been found to be problematic for rural level operations in developing countries. Moreover this also increases the operational cost of the system (Dysart, 2008);

- Less purified AA products have relatively low fluoride removal capacity (Fawell at al., 2006; Meenakshi et al., 2006);

- Spent regeneration solution contains high F concentrations (Feenstra et al., 2007; Fawell at al., 2006) and is difficult to dispose off.

1.4.1.2 Bone charcoal defluoridation method

Bone charcoal (Fig. 1.5) is a blackish, porous, granular material used as filter media for defluoridation. It has been described as the oldest known defluoridation agent and was used in the USA in the 1940s through to the 1960s. It is reported as working very well in Thailand and in some African countries (e.g Tanzania) at the domestic level (Dysart, 2008; Feenstra et al., 2007; Fawell at al., 2006).

Fig 1.6 Bone charcoal particles

Bone charcoal is prepared from animal bones by heating in specially designed kilns at temperatures ranging from 400 to 500 °C at controlled air supply. The heat treatment process removes organic materials (fats, oils and meat remains) from the bones and also activates or improves its fluoride removal capacity (Mjemgera and Mkongo, 2002). The major components of bone charcoal are calcium phosphate (57-80%), calcium carbonate (6-10%), and activated carbon (7-10%) (Feenstra et al., 2007; Fawell at al., 2006).

When in contact with water, bone charcoal is able to remove a wide range of pollutants (colour, taste, odour components), and has a specific ability for the up-take of fluoride from solution. The removal capability is thought to be due to its chemical composition, mainly as hydroxyapatite where one or two of the hydroxyl groups can be replaced by fluoride (Dysart, 2008; Fawell et al., 2006). In addition, it is also thought that the fluoride removal can be due to the reaction between calcium phosphate and fluoride or the replacement of the carbonate by fluoride to form an insoluble fluoroapatite. According to Fawell et al. (2006) the principal reaction for fluoride removal can be represented as:

$$Ca_{10}(PO_4)_6(OH)_2 + 2F^- \rightarrow Ca_{10}(PO_4)_6F_2 + 2OH^- \qquad\qquad (1.1)$$

The appropriate preparation of the bone charcoal is crucial to optimizing its properties as a fluoride removal agent (Dysart, 2008; Feenstra et al., 2007; Fawell et al., 2006; Negendra, 2003).

Limitations of the bone charcoal defluoridation method have been found to include the following:

- Due to local taboos and religious beliefs, the method is culturally not acceptable in many societies (e.g among Muslims, Hindus etc) as bone is used in the preparation of the filter media (Fawell et al., 2006; Dahi, 1996).
- The process depends on the local availability of adequate quantities of bones as raw material for the preparation of the media (Dysart, 2008; Fawell et al., 2006)
- If the bone charring process is not properly done the resulting filter media will have low fluoride removal capacity (Fawell et al., 2006)
- When water is treated with a poorly prepared bone charcoal media, it may taste and smell like rotten meat and may also be aesthetically not acceptable. Any occurrence of such taste and odour problems may put consumers off and could result in a total

rejection of the bone charcoal treatment method (Dysart, 2008; Feenstra et al., 2007; Fawell et al., 2006; Dahi, 1996).

1.4.2 Coagulation-flocculation-filtration process: The Nalgonda technique

The Nalgonda Technique is a defluoridation method adapted and developed by the National Environmental Engineering Research Institute (NEERI) in India, and named after the village where it was first pioneered. The technique has been described as probably the best known and most established fluoride removal method. It is the most widely used defluoridation method in India (Feenstra et al., 2007; Iyengar, 2003).

The Nalgonda method is an aluminum sulfate-based coagulation- precipitation-sedimentation-filtration process. In the Nalgonda technique, alum ($Al_2(SO_4)_3$. $18H_2O$) is added as a coagulant to the fluoride-contaminated water under efficient mixing conditions to ensure complete mixing. This induces the development of insoluble aluminum hydroxide ($Al(OH)_3$ micro-flocs which gather together into large settleable flocs. The removal of fluoride is accomplished by the adsorption of the negatively charged fluoride ions (F^-) in solution onto the aluminum hydroxide particles by electrostatic attraction and are subsequently separated from the water by sedimentation and filtration (Dysart, 2008; Fawell at al., 2006). When alum is added to water, the resulting solution becomes acidic. Simultaneous addition of lime ($Ca(OH)_2$) is required for pH adjustment to ensure a neutral pH (Dysart, 2008; Fawell at al., 2006; Meenakshi et al., 2006).

The various reaction which occur in the Nalgonda defluoridation process are presented by equations (1.2) to (1.5) (Dysart, 2008; Fawell at al., 2006).

Dissolution of Alum:

$$Al_2(SO_4)_3.\ 18H_2O \rightarrow 2Al^{3+} + 3SO_4^{2-} + 18H_2O \qquad (1.2)$$

Precipitation of aluminum hydroxide:

$$2Al^{3+} + 6H_2O \rightarrow 2\ Al(OH)_3 + 6H^+ \qquad\qquad (1.3)$$

Co-precipitation of fluoride:

$$F^- + Al(OH)_3 \rightarrow Al\text{-}F\ complex + undefined\ product \qquad\qquad (1.4)$$

pH adjustment:

$$6Ca(OH)_2 + 12H^+ \rightarrow 6Ca^{2+} + 12\ H_2O \qquad\qquad (1.5)$$

Some limitations/short comings of the Nalgonda defluoridation technique include:

- The efficiency of the technique is limited to about 70%. It is therefore not suitable in situations where the fluoride concentration in raw water is very high (Feenstra et al., 2007; Fawell et al., 2006).
- The operational costs of the Nalgonda technique are high (Meenakshi et al., 2006).
- The process requires correct dosing of chemicals, regular attendance and close monitoring to ensure effective fluoride removal. The labour, skills and time requirements on a sustainable basis, has been noted by UNICEF to be problematic at the rural community level (Dysart, 2008; Fawell et al., 2006).
- It produces sludge which is toxic and require to be disposed of safely.
- The use of aluminium sulphate as coagulant increases the sulphate ion concentration and can lead to cathartic effects (i.e. accelerated defeacation) (Meenakshi et al., 2006)
- A large dose of aluminium sulphate may be required for the process and it may get to a point where users complain of the taste of the treated water (Fawell et al., 2006). This can result in the users either by-passing the treatment unit and using the raw water directly or returning to their traditional and contaminated water sources.
- According to Meenakshi et al. (2006), excess residual aluminum in treated water, which could result from the use of aluminium sulphate in the technique can cause dangerous dementia disease as well as neurobehavioral, structural and biochemical

changes. It can also affect respiratory, cardiovascular, endocrine and reproductive systems. Excess aluminum in drinking has also been linked to the causes of Alzheimer's disease (Okuda et al. 2009; Sarkar et al., 2007; Sajidu et al., 2006).

1.4.3 Contact precipitation process

In this technique fluoride removal is accomplished by the addition of calcium chloride and sodium dihydrogenphospate (MSP) or "monosodium phosphate" to the raw water and then bringing the solution in contact with a bone charcoal medium already saturated with fluoride. The technique is reported as having high fluoride removal efficiency (Feenstra et al., 2007; Fawell et al., 2006; Dahi, 1996).

In a solution containing calcium, phosphate and fluoride, the precipitation of calcium fluoride and/or fluoroapatite is theoretically feasible. The precipitation process is however, practically impossible due to the slow nature of the reaction kinetics. Though the mechanism of the fluoride removal by the contact precipitation process is not fully understood, it is thought that the saturated bone charcoal acts as a catalyst for the calcium fluoride and/or fluoroapatite precipitation process. It also acts a filter media for the resulting precipitate, thereby accomplishing the fluoride removal (Fawell et al., 2006; Dahi, 1996).

According to Fawell et al. (2006), the reactions which occur in the contact precipitation defluoridation process can be represented as follows:

Dissolution of calcium chloride:

$$CaCl_2 2H_2O(s) + 2H_2O \rightarrow Ca^{2+} + 2Cl^- + 2H_2O \tag{1.6}$$

Dissolution of monosodium phosphate:

$$NaH_2PO_4\, H_2O(s) \rightarrow PO_4^{3-} + Na^+ + 2H^+ + H_2O \tag{1.7}$$

Precipitation of fluoroapatite:

$$10\, Ca^{2+} + 6\, PO_4^- + 2F^- \rightarrow Ca_{10}(PO_4)_6F_2(s). \tag{1.8}$$

Precipitation of calcium fluoride:

$$Ca^{2+} + 2F^- \rightarrow CaF_2(s) \hspace{5cm} (1.9)$$

The saturated bone char coal acts as a catalyst in the above reactions.

The limitations of the contact precipitation process include:

- Chemicals and trained staff are required for the process, which is often problematic and unsustainable if the process is to be used for water defluoridation in rural communities in developing countries.
- The process is new and still under investigation. It has so far been implemented only at the domestic level in Kenya and Tanzania (Eawag, 2009; IRC, 2007).
- The use of bone char as contact bed also has similar limitations/shortcomings as indicated for the bone charcoal defluoridation method (Feenstra et al., 2007; Fawell et al., 2006; Dahi, 1996)

1.4.4 Membrane filtration processes: Reverse osmosis (RO)

Reverse osmosis process is a type of membrane separation technology which has been noted to be highly effective for fluoride removal. It has a fluoride removal performance of 85-95%, and has been recommended by the US-EPA as one of the best available techniques for fluoride removal (Dysart, 2008; Feenstra et al., 2007).

In the reverse osmosis process, pressure is applied to force feed water through a semi-permeable membrane thereby removing dissolved solutes (including fluoride) through rejection by the membrane, based on particle size and electrical charges. The technique is a physical process and the reverse of natural osmosis as a result of pressure applied to the concentrated side of a membrane to overcome osmotic pressure (Dysart, 2008; Feenstra et al., 2007; Meenakshi et al., 2006)

The reverse osmosis technique is also noted to have the following limitations:

- The process is complex for rural water supply in developing counties: it require special equipment, a continuous supply of electricity and, skilled personnel which are mostly not available in rural areas of developing countrie (Dysart, 2008).

- High capital, operation and maintenance cost (Meenakshi et al., 2006).

- There is a significant loss of water (typically 10-35%) by the process which can have implications on groundwater resources and any water conservation measures (Feenstra et al., 2007; Meenakshi et al., 2006).

- The process also removes most ions present in the water though some are required as essential minerals for the human body (Dysart, 2008; Feenstra et al., 2007; Meenakshi et al., 2006).

- Disposal of concentrate with rejected contaminants (fluoride) is difficult.

1.5   Aim and scope of the study

In rural communities of Ghana, groundwater is the main source of water for drinking as the problems of managing surface water systems, which require treatment are usually beyond their capabilities. Also in Ghana, the rural communities are mostly small and scattered over large number of settlements. Under those conditions, groundwater is considered the realistic and most economic source of potable water, as the underlying aquifers can be tapped close to the demand centers in response to the dispersed nature of the rural settlements (MacDonalds, 2009; Gyau-Boakye and Dapaah-Siakwan, 1999). Unfortunately excess fluoride at concentrations beyond the WHO and Ghana Standards Board (GSB) guideline value of 1.5 mg/l (GSB, 1997; WHO, 2011), was noticed in the groundwater in the Northern region of Ghana in 2001 (CWSA-NR, 2007). As a result many otherwise successfully drilled boreholes (groundwater wells) have been closed down for human consumption (when the water quality analysis indicates the presence of fluoride contamination). This represent an economic loss to the country (i.e cost of borehole drlliing, including related/overhead cost). In addition to the economic loss, the situation also hampers efforts by the responsible water agencies in increasing access to potable water to rural communities and small towns in the Northern region of Ghana. Moreover, typical of many developing countries, the region is characterized by the incidence of many water-related diseases, including diarrhea and cholera. The incidence of these otherwise preventable diseases hampers productivity and hence the socio-economic development of the region. Thus the inability to accelerate the provision of safe water to some of the populace due to the presence of excess fluoride in the ground water source remains a major concern. A key to a better management of the groundwater resources in the Northern Region of Ghana is to study the occurrence of fluoride and its distribution in the groundwater in the region.

Due to the negative health impacts of excess fluoride in drinking water on the populace mostly in developing countries, the search for low-cost and appropriate technology for the removal of fluoride from drinking water remains very critical. Among the fluoride removal techniques, the adsorption process is widely considered the most appropriate, particularly for small community water source defluoridaton. This is due to its many advantages including flexibility and simplicity of design, relative ease of operation and cost-effectiveness as well as its applicability for contaminant removal even at low concentrations. The appropriateness of the adsorption process, however, largely depends on availability of a suitable adsorbent (Chen et al., 2010; Maliyekkai et al., 2006; Sakar et at., 2006; Ruthven, 1984). An adsorbent may be considered suitable, if it has a good practical and economic viability for a sustainable use for groundwater defluoridation, particularly in developing countries. Such an adsorbent may possess some charateristics including; high adsorption capacity, local availability in abundant quantities or, amenable for local production using indigenously available materials, low-cost of investment as well as operation and maintenance of adsorption treatment system, ease of handling, does not deteriorate in performance or quality with time hence amenable for long-term storage, potential for regeneration for reuse and potential for a simple safe disposal of spent adsorbent (Saha et al, 2009).

Several materials have been tested for water defluoridation including; manganese-oxide coated alumina, bone charcoal, fired clay chips, fly ash, calcite, sodium exchanged montimorillonite-Na^+, ceramic adsorbent, laterite, unmodified pumice, bauxite, zeolites, fluorspar, iron-oxide coated sand, calcite, activated quartz and activated carbon (Chen et al., 2011; Maliyekkai et al., 2006; Sarkar et at., 2006; Bhargava et al., 1992; Fan et al., 2003; Milo et al., 2010, Moges et al., 1996; Das et al., 2005; Mohapatra et al., 2004; Sujana and Anand, 2011; Laveccehia et al., 2012; Mahvi et al., 2012; Malakootian et al., 2011, Sun et al., 2011). Some of these adsorbents have shown certain degrees of fluoride adsorption capacity, however, the applicability of most is limited either due to lack of socio-cultural acceptance, inadequate capacity for practical use, high cost or their effectiveness only in extreme pH conditions. For example activated carbon is found to be effective at pH less than 3.0. This may require pH adjustment and consequently additional capital, operation and maintenance cost, and could limit feasibility of such a fluoride removal technology in remote rural areas of developing countries where more often than not the populace depend upon groundwater point sources normally with a pH range of 6.0 - 8.5 (Fan et al., 2003; Nigussie et al., 2007). Moreover some studied materials are either of fine particles or powders, which makes separation of adsorbent from aqueous solutions difficult after application. Such materials could also cause clogging and/or low hydraulic conductivities, when applied in fixed-bed adsorption systems (Han et al., 2009).

The most acceptable and commonly used adsorbent is activated alumina (AA) which is considered the industry standard for drinking water defluoridation in developed countries. Activated alumina is, however, generally both expensive and not readily available in most developing countries, its regeneration and disposal of the exhausted adsorbent are major challenges (Maliyekkal et al., 2006).

For sustainable drinking water defluoridation particularly in developing countries, however, the search for appropriate fluoride adsorbents with a potential for local in-country production continue to remain of crucial interest.

This study focused on gaining an insight into the occurrence of fluoride in groundwater in the Northern region of Ghana. The study further explored the use of indigenous materials as base material for the synthesis of fluoride adsorbent which could have the potential for local production, hence a sustainable application for drinking water defluoridation.

The fluoride adsorption performance of the synthesized adsorbents were studied in laboratory batch and column experiments. Various characterization techniques and, several kinetic and equilibrium isotherm models were employed to understand the physico-chemical characteristics and fluoride adsorption properties/capacities of the synthesized adsorbent, as well as the mechanism of the fluoride removal. The overall goal of the study was therefore twofold: (i) to study the groundwater chemistry in the Northern region of Ghana with focus on the occurrence, genesis and distribution of fluoride-contaminated waters in the eastern corridor of the region (which is the most fluoritic part), and, (ii) to contribute to the search for an appropriate and sustainable fluoride removal technology for the treatment of the fluoride-contaminated groundwater for drinking water production in developing countries.

1.6   Research objectives

The specific objectives of the research were:

(i)   To investigate the factors controlling the occurrence of fluoride and its distribution in groundwaters in the Northern region of Ghana, with a focus on the Eastern corridor which is the most affected area.

(ii)   To explore the possibility of synthezing fluoride adsorbents using locally available base materials (pumice, bauxite and charcoal) for sustainable drinking water defluoridation.

(iii)   To conduct kinetic, equilibrium isotherm and column experiments using the synthesized adsorbents, estimate their adsorption capacities and compare against the industry standard, activated alumina (AA).

(iv)   To study the impact of pH and co-ions on the fluoride removal behavior of the adsorbents.

(v)   To characterize the selected adsorbents to determine their physico-chemical characteristics with focus on their mineralogy, surface chemistry, specific surface area and porosity.

(vi)   To determine the proposed mechanisms responsible for fluoride removal onto the selected adsorbents.

(vii) To screen the feasibility of different options for simple regeneration of exhausted adsorbents.

(viii) To assess different options for fixing/stabilizing of fluoride saturated adsorbent that could permit safe disposal in a land fill.

(ix)   To fit adsorption data to (existing) kinetic and equilibrium models to describe the adsorption of fluoride onto the selected adsorbent.

(x)   To assess the performance of the most promising adsorbent for fluoride removal in fluoritic areas in Northern Ghana.

1.7 Outline of the thesis

This thesis is made up of 7 chapters: five of the chapters represent the results and findings of different segments of the research, in addition to an introductory and conclusion chapters.

After this introductory chapter, **chapter 2**, presents an examination of the occurrence, distribution and genesis of high fluoride in groundwater in the Northern Region of Ghana with a focus on the Eastern corridor of the region. The chapter discusses proposed mechanisms responsible for the mobilization of fluoride into the ground water system of the study area.

The next four chapters (3, 4, 5 & 6) present results from laboratory studies on the synthesis of fluoride adsorbents for water defluoridation, that could have a potential for sustainable application in fluoritic areas of developing countries, based on surface modifications of indigenous/locally available materials (pumice, bauxite and wood charcoal) as base material.

Thus **Chapter 3** focuses on surface modification by an Al coating process using pumice as base and testing the efficacy of the synthesized adsorbent, aluminum oxide coated pumice (AOCP) for fluoride removal in batch adsorption experiments. The hard soft acid base (HSAB) concept, which was explored in the surface modification process is discussed. The chapter further discusses the mechanism of fluoride adsorption onto AOCP.

Chapter 4 assesses the efficacy of AOCP for water defluoridation under continuous flow conditions in laboratory- scale column experiments, and the modeling of breakthrough. The chapter also presents findings from studies on adsorbent regeneration and evaluates options for safe disposal of spent adsorbent into the environment.

Chapter 5 presents work on adsorbent systhesis by aluminum coating, using raw bauxite as base material, with a focus on investigating the effects of various systhesis process conditions on the defluoridation efficiency of the surface modified bauxite. Results on the effect of co-ions commonly found in groundwater on the efficacy of the coated bauxite are presented.

Chapter 6 focuses on surface treatment/functionalization of wood charcoal, which is the cheapest and most readily available of the three indigenous base materials, in most developing countries. The chapter presents results on batch and continuous flow column experiments for examining the efficacy of aluminol functionalized wood charcoal (AFWC) for fluoride removal. In addition to laboratory studies, the chapter also presents findings of field testing of AFWC for house-hold level treatment of fluoride-contaminated groundwater in Bongo town in the Upper East region of Ghana, where excess fluoride in the drinking water has adverse health and social impacts on most of the populace. Results from similar field test conducted using activated alumina (AA), the standard industrial fluoride adsorbent for comparison to AFWC, are also presented. And finally, **Chapter 7**

provides the overall conclusions, outlook and recommendations for further research.

References

Apambire, W. B., Boyle, D. R., Michel, F. A. 1997. Geochemistry, genesis, and health implications of fluoriferous groundwaters in the upper regions of Ghana. Environ. Geology, 33 (1), 13-24.

Attanayake, M.A.M.S.L., Padmasiri, J.P., Fernando, W.S.C.A. 1995. Sustainability of water and sanitation systems: Laterite for water treatment. 21st WEDC Conference, Kampala, Uganda.

Ayamsegna, J.A., Apambire, W.B., Bakobie, N., Minyila, S.A. 2008. Removal of fluoride from rural drinking water sources using geomaterials from Ghana. 33rd WEDC International Conference, Accra.

BGS. 2009. Water quality fact sheet: Fluoride, British Geological Survey

Bhargava, D.S., and Killedar, D.J. 1992. Fluoride adsorption on fishbone charcoal through a moving media adsorber, Water Res. 26(6), 781-788.

Biswas, K., Bandhoyapadhyay, D., Ghosh, U.C. 2007. Adsorption kinetics of fluoride on iron (III) zirconium hybrid oxide. Adsorption, 13, 83-94.

Boddu, V.M., Abburi, K., Talbott, J.L., Smith, E.D., Haasach, R. 2008. Removal of arsenic (III) and arsenic (V) from aqueous medium using chitosan-coated biosorbent. Water Res, 42: 633-642.

Buamah, R., Petrusevescski, B., Schippers, J.C. 2008. Adsorptive removal of manganese (II) from the aqueous phase using iron oxide coated sand. J. of Water Supply and Tech., Aqua, 57(1), 1-11

Chen, N., Zhang, Z., Feng, C., Suguira, N., Li, M., Chen, R. 2010. Fluoride removal from water by granular ceramic adsorption, J. Coll. Interf. Sci. 348, 579-584.

Chen, N., Zhang, Z., Feng, C., Zhu, D., Yang, Y., Sugiura, N. 2011. Preparation and characterization of porous granular ceramic dispersed aluminum and iron oxides as adsorbent for fluoride removal from aqueous solution, J. of Hazard. Mater. 186, 863-868.

CWSA-NR. 2007. Summary of the fluoride issues in the northern region: Report submitted to the Government of Ghana. Community Water and Sanitation Agency, June 2007, Tamale, Northern region.

Dahi, E. 1996. Contact precipitation for defluoridation of water. Paper presented at 22nd WEDC Conference, New Delhi, India, 9-13 September.

Das, N., Pattanaik, P., Das, R. 2005. Defluoridation of drinking water using activated titanium rich bauxite, J. of Coll. and Inter. Sci. 292, 1- 10.

Dysart, A. 2008. Investigation of defluoridation options for rural and remote commmunities. Report No.41. Cooperative Research Centre for Water Quality and Treatment, Salisbury, Australia.

Eawag. 2009. Fluoride removal from drinking water, Swiss Federal Institute of Aquatic ScienceandTechnology.
http://www.wrq.eawag.ch/activities/treatment/fluoridetreatment/index_EN. Cited October 2009.

Fan, X., Parker, D.J., Smith, M.D. 2003. Adsorption kinetics of low cost materials, Water Res.37, 4929-4937.Fawell, J., Bailey, K., Chilton, J., Dahi, E., Fewtrell, L., Magara, Y. 2006. Fluoride in drinking water. IWA Publishing, London, 4-81.

Feenstra, L., Vasak, L., Griffion, J. 2007. Fluoride in groundwater: Overview and evaluation of treatment methods. igrac International groundwater Resources Assessment Centre. http://www.igrac.net/dynamics/modules/FIL0100/view.php?fil_ld=131.

Firempong, C.K., Nsia, K., Awunyo-Vitor, D., Dongsogo, J. 2013. Soluble fluoride levels in drinking water: A major risk factor of dental fluorosis among children in Bongo communiyu of Ghana, GhanaMedical J. 47 (1), 16 – 23.

Gessner, B.D., Beller, M., Middaugh, J. P., Whitford, G.M. 1994. Acute fluoride poisoning from public water system. The New England J. of Medicine. 330 (2), 95 – 99.

GSB. 1997. Water quality: Requirements for drinking water, Ghana Standard Boards, GS 175 PT.1:1997, Accra, Ghana.

Gyau-Boakye, P and Dapaah-Siakwan, S. 1999. Groundwater: solution to Ghana's rural water supply industry? *The Ghana Engineer* http://home.att.net/¬africantech/GhIE/ruralwtr/ruralwtr.htm. Cited June 2009.

Han, Y., Park, S., Lee, C., Park, J., Choi, N., S. Kim, S. 2009. Phosphate removal from aqueous solution by aluminum (hydr) oxide-coated sand. Korean Society of Env. Eng.14 (3), 164-169.

Harder, R. 2008. Fluoride-Toxin or medicine. Department of Water Management, Delft University of Technology, Delft, Netherlands.

Hurtado, R., Gardea-Torresdey, J., Teiman, K .J. 2000. Fluoride occurrence in tap water at Lod altos de Jalisco in the Central Mexico region. Proceedings of the 2000 Conference on Hazardouswaste Research. http://www.engg.ksu.edu?HSRC/ooProceed/gardea_torredey1.pdf. Cited September 2009.

IRC. 2007. Household water treatment: FAQ sheet. http://www.irc.nl/page/8028. Cited October 2009

Iyengar, L. 2003. Defluoridation of water using activated alumina technology. Final report submitted to UNICEF. Department of Chemistry, Indian Institute of Technology, Kanpu, India.

Ku, Y and Chiou, H. 2002. The adsorption of fluoride from aqueous soluteon by activated alumina. Water, Air and Soil pollution, 133, 349-360.

Lavecchia, R., Medici, F., Piga, L., Rinaldi, G., Zuorro, A. 2012. Fluoride removal from water by adsorption on a high alumina content bauxite, Chem. Eng. Trans., 26, part 1, 225-230.

Lu, Y., Su, Z.R., Wu, L.N., Wang, X., Lu, W., Liu, S.S. 2000. Effect of high-fluoride water on intelligence in children. Fluoride, 33 (2) 74 -78.

Ma, W., Ya, F., Han, M., Wang, R. 2007. Characteristics of equilibrium, kinetics studies for adsorption of fluoride on magnetic-chitosan particle. J. of Hazard. Mat., 143, 296-302.

MacDonald, A. 2009. Groundwater and rural water supply in Africa. IAH Burdon GroundwaterNetwork. http://wwww.iah.org/downloads/occpub/IAA_ruralwater.pdf. Cited May 2009

MacDonald, A.M and Davis, J. 2000. A brief review of groundwater for rural water supply in Subsaharan Africa. BGS Technical Report WC/00/33. http://www.edsworldbank.org/eds/ard/grounwater/pdfreports/reviewGdwtrSubs aharanAfrica pt1.pdf. Cited May, 2009.

Madhuure, P., Sirsikar, D.Y., Tiwari, A.N., Ranjan,B., Malpe,D.B. 2007. Occurrence of fluoride in the groundwaters of Pandharkawada area, Yavatmal district, Maharashtra, India. Current Science, 92(5), 675-678.

Mahvi, A.H., Heibati, B., Mesdaghinia, A., Yari, A.Y. 2012. Fluoride adsorption by pumice from aqueous solution, E-Journal of Chem. 9(4), 1843 -1853.

Malakootian, M., Moosazadeh, M., Yousefi, N., Fatehizadeh, A. 2011. Fluoride removal from aqueous solutions by pumice: case study on Kuhbonan water, African J. of Environ. Sci. and Tech., 5(4), 299-306.

Maliyekkal S M, Sharma A K, Philip L. 2006. Manganese-oxide-coated alumina: A promising sorbent for defluoridation. Water Res., 40, 3497-3506.

Meenakshi, R.C., Maheshwari. 2006. Fluoride in drinking water and its removal. Center for Rural Development and Technology. Indian Institute of Technology.

Mjemgera, H., Mkongo, G. 2002. Appropriate defluoridation technology for use in fluoritic areas in Tanzania: Water demand for sustainable development. 3rd WaterNet/Warfa Symposium, Dar es Salaam, Tanzania, 30-31st October 2002.

Mjengera, H.J., Mkongo, G.B. 2009. Occurrence of fluoride in water sources and water defluoridation in Tanzania. Ministry of Water, Dar es Salaam, Tanzania.

Mlilo, T., Brunson, L., Sabatini, D. 2010. Arsenic and fluoride removal using simple materials, J. Environ. Eng. 136(4), 391-398.

Moges, G., Zewge, F., Socher, M. 1999. Preliminary investigations on the defluoridation of water using fired clay chips, J. of African Earth Sci. 21(4), 479-482.

Mohapatra, D., Mishra, D., Mishra, S.P., Chaudhury, G.R., Das, R.P. 2004. Use of oxide minerals to abate fluoride from water, J. of Coll. and Interf. Sci. 275, 355 - 359.

Negendra, C. R. 2003. Fluoride and Environment—A Review. Third international conference on environment and health, Chennai, India, 15-17 December, 2003. http://www.uorku.ca/bunchmj/ICEH/proceeding/rao_N_ICE_papers_386to399. pdf.

Nigussie, W., Zewge, F., Chandravanshi, B.S. 2007. Removal of excess fluoride from water using waste residue from alum manufacturing process, J. of Hazard. Mater.147, 954-963.

Okuda, T., Base, A.U., Nishijima, W., Okada, M. 2009. Improvement of extraction method of coagulation active components from *Moringa oleifera* seed. Department of Environmental Science, Hiroshima University, Hiroshima, Japan.

Petrusevski, P., van der Meer, W., Baker, J., Kruis, F., Sharma, S.K., Schippers, and J.C. 2007. Innovative approach for treatment of arsenic contaminated groundwater in Central Europe. Water Science & technology: *Water Supply,* **7** (3), 131 -138.

Pollard, S.J.T., Thompson, F.E., McConnachie, G.L. 1994. Microporous carbons from Moringa Oleifera husk for water purication in less developed countries. Water Res., 29(1), 337-347

Rocha-Amador, D., Navarro, M.E., Carrizales, L., Morales, R., Calderon, J. 2007. Decreased intelligence in children and exposure to fluoride and arsenic in drinking water. Cad. Saude Publica, Rio de Janeiro, 23 Sup 4: S579 – S587.

Ruthven, D.M. 1984. Principles of adsorption and adsorption processes. John Wiley & Sons, Inc, USA.

Sahli, M.A.M., Annour, S., Tahaikl, M., Mountadar, M., Soufiane, A., Elmidaoui, A. 2007. Fluoride removal for underground brackish water by adsorption on the natural chitosan and by electrodialysis. Desalination, 212, 37-45.

Sajidu, S.M., Henry, E.M.T., Persson, I., Masamba, W.R.L., Kayambazinthu, D. 2006. pH dependence of sorption of Cd^{2+}, Zn^{2+}, Cu^{2+} and Cr^{3+} on crude water and sodium chloride extracts of Moringa stenopetala and Moringa oleifera. African J. of Biotechnol., 5(23), 2397-2401.

Salifu, A., Petrusevski, B., Ghebremichael, K., Buamah, R., G. Amy, G. 2012. Multivariate statistical analysis for fluoride occurrence in groundwater in the Northern region of Ghana J. Contaminant Hydro. 140–141, 34–44.

Sarkar, M., Banerjee, A., Pramanick, P.P., Sarkar, A.R. 2007. Design an operation of fixed bed laterite column for the removal of fluoride from water. Chem. Eng. J., 131, 329-335.

Shomar, B., Muller, G., Yahya, A., Aska, S., Sansur, R. 2004. Fluoride in groundwater, soil and infused black tea and the occurrence of dental fluorosis among school children of the Gaza strip. J. of Water and Health, 2 (1), 23-35.

Shin, R.D. 2016. Fluoride toxicity: Background, Pathophysiology, Etiology. http://emedicine.medscape.com/article/814774-overview.

Sujana, M.G and Anand, S. 2011. Fluoride removal studies from contaminated groundwater by using bauxite, Desalination. 267, 222-227.

Sun, Y., Fang, Q., Dong, J., Cheng, X., Xu, J. 2011. Removal of fluoride from drinking water by natural stilbite zeolite modified with Fe (III), Desalination. 277, 121-127

UN General Assembly. 2010. Resolution adapted by the General Assembly on July 2010, A/RES/64/292: The human right to water and sanitation. Meetings coverage and Press release.

UNICEF & WHO. 2015. 25 years Progress on sanitation and water: 2015 update and MDG assessment, WHO, Geneva, Switzerland.

UNICEF, 2009a. Water, Sanitation and Hygiene. http://www.unicef.org/wash. Cited August, 2009.

UNICEF, 2009b. UNICEF's Position on water fluoridation. Water environment and sanitation. http://www.nofluoride.com/UNICEF_fluor.htm. Cited August, 2009.

UNICEF. 2004. High fluoride in drinking water: An overview, International Conference and Exhibition on Groundwater, Addis Ababa, 25-27 May, 2004.

UNICE. 2001. Domestic defluoridation: Community effort for making drinking water safe, Ways to watsan. Rajasthan, India.

Valdez-Jimenez, L., Fregozo, C.S., Beltran, M.L.M., Coronnado, O. G., Vega, M.I.P. 2010. Effects of the fluoride on the central nervous system. Sociedad Espanola de Neurologia. Elsevier Espania, S.L.

Walther, I. 2009. Fluoride contamination in the main Ethiopian Rift Valley. http://www.geo.tufriberg.de/fernerkundung/PAPERS/fluor_walther.pdf. Cited September, 2009.

Whitford, G.M. 1996. The metabolism and toxicity of fluoride, Karger, Switzerland.

WHO, 2011. Guideline for Drinking Water Quality Incorporating. Forth edition. World Health Organization, Geneva.

Yu, Y., Yang, W., Dong, Z., Wan, C., Zhang, J., Liu, J., Xiao, K., Huang, Y., Lu, B. 2008. Neurotransmitter and receptor changes in the brains of foetuses from areas of endemic fluorosis. Fluoride, 41 (2), 134 – 138.

Zhao L.B., Liang, G.H., Zhang, D.N., Wu, X.R. 1996. Effect of a high fluoride water supply on children's intelligence. Fluoride, 29, 190-192.

2

Fluoride occurrence in groundwater in the Northern region of Ghana

Main part of this chapter was published as:

Salifu, A., Petrusevski, B., Ghebremichael, K., Buamah, R., Amy, G.L. 2012. Multivariate statistical analysis for fluoride occurrence in groundwater in the Northern Region of Ghana. J. of Contaminant Hydro. 140-14, 34-44.

Abstract

The presence of fluoride in groundwater in parts of the Northern region of Ghana at levels above 1.5 mg/L, the WHO guideline, value has exposed the population in the fluoritic communities to fluoride-related health hazards. Piper graphical classification, principal component analysis (PCA) and thermodynamic calculations were used as an approach to gain insight into the groundwater chemical composition and to help understand the dominant mechanisms influencing the occurrence of high fluoride waters. Inverse distance weighting interpolation (IDW) and spatial join procedure were used to examine the relationship between the underlying geology of the study area and fluoride distribution. The fluoride concentration in 357 groundwater samples from the area ranged between 0.0 and 11.6 mg/L, with a mean value of 1.13 mg/L. Six groundwater types were identified for the area: Ca-Mg-HCO$_3$, Ca-Mg-SO$_4$, Na-Cl, Na-SO$_4$, Na-HCO$_3$ and mixed water type. PCA performed on the groundwater chemical data resulted in 4 principal components (PCs) explaining 72% of the data variance. The PCs represented the predominant processes controlling the groundwater chemistry in the study area which include, mineral dissolution reactions, ion exchange processes and evapotranspiration processes. PHREEQC calculations for saturation indices of the groundwater samples indicated they were largely saturated with respect to calcite and under-saturated with respect to fluorite, suggesting that dissolution of fluorite may be occurring in the areas where it is present. A review of the PCA results and an evaluation of the equilibrium state of the groundwater based on the saturation indices suggest that the processes controlling the overall groundwater chemistry in the area also influenced the fluoride enrichment. This predominant processes include the dissolution of the mineral fluorite, anion exchange processes (F$^-$/OH$^-$) involving clay minerals and evapotranspiration processes. Elevated fluoride levels in the study area were found to occur predominantly in the Saboba and Cheriponi districts and also in the Yendi, Nanumba North and South districts. These areas are underlain by the Obossom and Oti beds, comprising mainly of sandstone, limestone, conglomerate, shale, arkose and mudstone. Results of the conducted hydrochemical analysis show that aside the boreholes with elevated concentrations of fluoride (beyond 1.5 mg/L), groundwater in the study area based on the parameters analysed is chemically acceptable and suitable for domestic use.

2.1   Background

The presence of fluoride in groundwater in parts of the Northern region of Ghana at levels above 1.5 mg/L, the WHO guideline value, has exposed the population in the fluoritic communities to fluoride-related health hazards. A limited fluorosis survey conducted in six selected communities in the eastern corridor of the Northern region of Ghana, where the presence of high fluoride in groundwater is most prominent, revealed the incidence of dental fluorosis (tooth mottling). The prevalence rate in some of the communities was found to be as high as 61%. The severe type of dental fluorosis based on Dean's fluorosis index classification criteria, which is characterised by brown stains, discrete or confluent pitting of the dental enamel was observed. The possible emergence of skeletal fluorosis (bone deformation and painful brittle joints in older people) was also observed in some of the affected communities (Mandinic et al., 2010; Ayugane, 2008).

The presence and distribution of elevated concentrations of fluoride in groundwater and the related health impacts in other parts of Ghana, particularly the Bongo and Bolgatanga districts in the Upper East region, have also been reported and well studied (Apambire et al., 1997; Anongura, 1995). The fluoride concentrations in the area were found to range between 0.11 and 4.60 mg/L. The fluoride enrichment was also found to be most probably associated with the dissolution of the mineral fluorite (CaF_2), found in the Bongo granites in the area, and also from both the dissolution and anion exchange from micaceous minerals and their altered clay products.

Even though groundwater remain the most important source (about 90 %) for rural water supply in the Northern region of Ghana, little is known about the natural and/or anthropogenic factors that control the groundwater chemistry, and hence the groundwater quality and fluoride contamination. The aim of this work was to study the groundwater chemistry in the Northern region with a focus on the occurrence, distribution and genesis of high fluoride waters in the eastern corridor of the region. Piper graphical classification, Pearson's correlation, principal component analysis (PCA) and thermodynamic calculations were used as an approach to gain an insight into the groundwater chemical composition and the dominant mechanisms influencing the occurrence of high fluoride waters. Inverse distance weighting interpolation (IDW) and spatial join procedure were used to examine the relationship between the underlying geology of the study area and fluoride distribution.

2.2   Occurrence and hydrogeochemistry of fluoride

Fluorine is the lightest member of the halogen group and one of the most reactive of all chemical elements. It is therefore not found as fluorine in the environment. Fluorine is also the most electronegative of all elements and therefore, has a strong tendency to acquire a negative charge to form a monovalent ion, fluoride (F-) (Harder, 2008; Fawell et al., 2006).

Fluorine is widely distributed in the earth's crust as the fluoride ion and constitutes about 0.06-0.09% weight of the upper layers of the lithosphere (Fawell et al., 2006; Weinstein & Davison, 2004). Fluoride and hydroxide ions both have the same charge and similar ionic radii, and can therefore replace each other in many mineral structures to form fluoride-mineral complexes. Over 150 minerals are known to contain fluoride but with a great variation in content from as high as 73% in the rare mineral griceite (LiF) to less than 0.2% in many others (Walther, 2009; Weinstein & Davison, 2004). The most common fluoride-bearing minerals, which are the natural sources of fluoride include; fluorite (CaF_2), fluoroapatite ($Ca_{10}(PO_4)_6F_2$), villianmite (NaF), cryolite (Na_3AlF_6), topaz ($Al_2SiO_4(F,OH)_2$), the micas including muscovite ($KAl_2(Si_23Al)O_{10}(OH,F)_2$) and biotite ($K(Mg,Fe)_3AlSi_3O_{10}(OH,F)_2$), the amphiboles including hornblende ($NaCa_2(Mg,Fe,Al)_3(Si,Al)_8O_{22}(OH,F)_2$) and rock phosphate. Clay minerals such as illite, chlorite and smectites also represent excellent anion exchange media that can contribute to fluoride enrichment of groundwater (Brunt et al., 2004; Weinstein & Davison, 2004; Apambire et al, 1997; Frencken et al., 1992). Due to its favorable dissolution properties, however, fluorite (CaF_2) seems to be the main mineral that predominantly controls aqueous fluoride geochemistry in most environments (Mamatha and Rao, 2010; Weinstein & Davison, 2004; Apambire et al., 1997).

Fluoride occurs in practically all natural groundwaters, in concentrations varying from trace to as high as 2,800 mg/L in environments such as the Soda Lakes of the East African Rift System (Tiemann, 2006; Hurtado et al., 2000; Apambire et al. 1999).

The dominant factors that control the concentration of fluoride in natural groundwater include: the geological setting and types of rocks/minerals traversed by the groundwater, solubility of fluorine-bearing minerals in the aquifer matrix and the amount of leachable fluorine they contain, anion exchange capacity of aquifer materials (OH⁻ for F⁻), and the chemical composition of the traversing groundwater and its contact time or reaction time with fluorine-bearing materials (Biswas et al., 2007; Brunt et al., 2004; Hurtado et al., 2000; Apambire et al. 1997). For instance, alkaline pH ranging from 7.6 to 8.6 with high HCO_3 concentration (350 - 450 mg/L) and moderate electrical conductivity (EC), are the favourable conditions for CaF_2 dissolution in groundwater and hence, the release of fluoride into groundwater system (Walther, 2009; Mjengera and Mkongo, 2009; Madhnure et al., 2007; Shomar et al, 2004; Nagendra, 2003; Hurtado et al., 2000; Apambire et al., 1997). The concentrations of Ca, Na, hydroxyl ion, and certain complexing ions such as Fe, Al, B, Si, if present in groundwater, can also affect its fluoride content (Apambire et al., 1997). For instance due to the common ion effect, the dissolution of fluorite is suppressed when the concentration of Ca is high, above the limit for fluorite solubility. This lowers the release of fluoride into the groundwater system resulting in low concentrations. On the other hand high-fluoride ground waters are mainly associated with sodium-bicarbonate water types and relatively low calcium and magnesium concentrations. Groundwater chemical composition can therefore be a useful proxy indicator of potential fluoride problems (Ayoob and Gupta, 2009; British Geological Survey, 2009; Biswa et al., 2007; Brunt et al., 2004).

The climatic conditions of an area of interest may also have an influence on the fluoride concentration of the groundwater (Biswas et al., 2007; Brunt et al., 2004; Hurtado et al., 2000; Apambire et al., 1997). Arid regions are more vulnerable to high fluoride concentrations. In these regions groundwater flow is slow and reaction times with rocks (including fluoride-bearing minerals) are therefore longer (British Geological Survey, 2009; Madhnure et al., 2007). Also high fluoride concentrations tend to occur in arid regions due to high evapotranspiration, which concentrates the fluoride. Dissolution of evaporative salts deposits in arid environments may also be an important source of fluoride (Brunt et al., 2004; Weinstein & Davison, 2004). On the other hand fluoride enrichment of groundwater is less pronounced in humid tropical regions because of high rainfall inputs and the diluting effect on groundwater composition (British Geological Survey, 2009; Brunt et al., 2004).

High fluoride concentrations can also build up in ground waters which have long residence times in host aquifers. Such groundwaters are usually associated with deep aquifer systems and slow groundwater movement. On the other hand shallow groundwaters, which represent young and recently infiltrated rain water generally have low concentrations of fluoride. Such groundwaters are usually found in hand dug wells (British Geological Survey, 2009; Madhnure et al., 2007).

Fluoride contamination of groundwater can also result from anthropogenic sources and/or industrial activities such as application of phosphatic fertilizers, processing of phosphatic raw materials, use of clays in ceramic industries, electroplating, aluminum smelting operations and burning of coal (Chen et al., 2010; Naani et al., 2008; Brunt et al., 2004; Kundu et al., 2001; Frencken, 1992).

2.3 The study area

The eastern corridor of the Northern region of Ghana (the study area) is located between latitudes 8° 30" and 10° 30" N and longitudes 1° 0" W and 1° 0" E. It comprises eleven administrative districts (West Mamprusi, East Mamprusi, Bunkpurugu Yunyo, Gushegu, Karaga, Saboba/Cheriponi, Zabzugu Tatale, Yendi, Nanumba North and South (Fig.2.1), and covers an area of about 27,900 km^2 with a population of about 1,150,000.

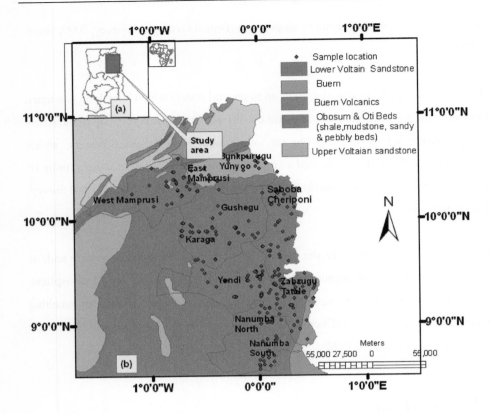

Fig. 2.1 The study area: (a) Ghana and (b) eastern corridor of Northern region of Ghana, indicating the geology of the area and locations of sampled boreholes.

Temperatures in the Northern region of Ghana are relatively high, ranging from a minimum of 12°C in January to a maximum of 43°C in April, with a mean annual value of 30°C. The climate of the area is classified as semi-arid (WRC, 2006; Asiamah et al., 1997), and the mean annual rainfall is around 900mm.

The study area is underlain largely by Neoproterozoic to Paleozoic sedimentary rocks, locally referred to as the Voltain sedimentary formation. The formation was developed in a depression of the underlying West African Craton, which consists mainly of crystalline igneous and metamorphic rocks of the Precambrian age. The Voltain formation has generally been divided into the Lower, Middle (Obossom and Oti beds) and Upper Voltain formations based on lithology and field relationships (Fig 2.1). The Middle Voltain formation is the largest of the Voltain formation. The Voltain sediments consist mainly of sandstone but also include shale, mudstone, arkoses, grey wacke and siltstone. Some limestone, conglomerate and evaporates also occur in the formation. The sediments originated mainly from a glacial event followed by prolonged marine incursions (WRC, 2006; Anani, 1999; Acheampong and Hess, 1998; Kesse, 1985; Gill, 1969; Junner and Hirst, 1946; Junner and Service, 1936). Parts of the eastern portion of the study area is underlain by the Buem formation, which is also dominated by shale, sandstone, lava and tuff with some limestone, grit and conglomerate.

The mineralogy of the sandstone, conglomerate and shale in the sediments is found to be dominated by sodic plagioclase feldspars and quartz with minor chlorite, calcite, talc, K-feldspar, zeolite, saponite and some micas including muscovite and biotite (WRC, 2006; Acheampong and Hess, 1998). The main minerals found in the limestone and dolomite are Ca -and Mg-carbonate. These carbonate minerals are also found in the sandstones and clays, either as accessory minerals, or as cement around the inert grains (Appelo and Postmas, 2005; Freeze and Cherry, 1997). Na-montmorillonite is the main clay mineral phase in the Voltain formation (Yidana, 2010). Minerals also found in environments with evaporates include calcite, dolomite, gypsum, anhydrite and halite (Hounslow, 1985).

Groundwater occurrence and flow in the formation is mainly controlled by the presence of secondary permeability and along bedding plains, due to the loss of primary porosity (Yidana, 2010; Gill, 1969; WRC, 2006).

2.4 Study methodology

Three hundred and fifty seven (357) groundwater samples taken from boreholes drilled in about 300 communities in the study area, were analysed for the chemical data using standard methods. The spatial distribution of the sampled boreholes is shown on Fig 2.1. The parameters analysed for each sample included pH, Temp.(^{0}C), electrical conductivity (EC), Total dissolved solids (TDS), Na^+, K^+, Ca^{2+}, Mg^{2+}, SO_4^{2-}, HCO_3^-, Cl^-, NO_3^-, F^- and total hardness. Unstable hydrochemical parameters such as EC, pH and alkalinity were measured in situ immediately after collection of samples, using a WTW LFT 91 field conductivity meter model, WTW 95 field pH meter and a HACH digital titrator, respectively, that had been calibrated before use. At each borehole site, the water was pumped for more than 10 minutes prior to sampling in order to get a representative sample. Hand-held syringes fitted with a filter head with 0.45μm cellulose filter membrane were used to filter the water samples in the field. Polyethylene sampling bottles (100 ml), washed with demineralised water were used for sample collection. Sampling bottles for cations were acidified with concentrated reagent grade HNO_3 acid. The bottles were tightly capped to protect samples from atmospheric CO_2, adequately labeled and preserved in a refrigerator until they were taken to the laboratory for measurement. Samples were analyzed both at the laboratories of the Water Research Institute in Tamale, Ghana and at the UNESCO-IHE analytical laboratory, Delft, the Netherlands.

Principal component analysis (PCA), a multivariate statistical technique was used to condense the multidimensional groundwater quality data into a lower dimensional format, and to provide an insight into the underlying hydrogeochemical as well as any anthropogenic processes possibly influencing the groundwater chemistry within the aquifers of the study area (Vlassopoulos et al., 2009; Chen et al., 2007; Praus, 2007). The statistical software package SPSS 18.0 for Windows was used for performing the PCA and for calculation of basic descriptive statistics. The number of principal components (PCs) considered from the PCA were determined on the basis of the Kaiser criterion of eigen values greater than or equal to 1 (Kaiser, 1960), and from a Cattel scree plot (Cattel, 1966). The variamax orthogonal rotation was used in order to facilitate the PCA interpretation. The ground water samples were plotted on a Piper diagram for classification using the GW Chart software. The equilibrium states of all the groundwater samples with respect to possible reactant and product minerals were evaluated through their saturation indices (SI). The saturation indices of minerals were subsequently used for determining which direction geochemical processes (water-rock reactions) within the aquifers may go and hence the influence on the groundwater chemistry (Appelo and Postmas, 2005; Rogers, 1989). Speciation calculations and calculation of mineral saturation indices, expressed as $SI = \log (IAP/K)$, for all the groundwater samples were carried out using PHREEQC for Windows, where IAP is the ion activity product of the dissociated chemical species in solution and, K is the equilibrium solubility product for the chemicals involved at the sampled temperature (Yidana et al., 2008; Parkhurst and Appelo, 1999). The relationship between the geology of the study area and fluoride distribution was examined by superimposing the geo-referenced groundwater chemical data onto a digitised geology map using spatial join procedure. Inverse distance weighting interpolation (IDW) was also applied to the spatial dataset to produce a prediction map for fluoride for the study area. The spatial analysis was performed using Arcgis 9.2.

2.5 Results and Discussions

2.5.1 Descriptive statistics

A univariate statistical analysis of the hydrochemical data of the groundwater samples collected from the area, which include the lower quartile (Q_1), median (Q_2), upper quartile (Q_3) and the interquartile range (IQR) are presented in Table 2.1.

Table 2.1: Descriptive statistics of groundwater chemical data

Parameter[a]	Min.	Max.	Mean	Std. deviation	Q_1	Q_2	Q_3	IQR
pH	6.10	9.53	7.80	0 .56	7.49	7.80	8.19	0.70
EC	8.53	5110.00	743.75	501.00	500.00	690.00	907.00	407.00
TDS	7.99	3372.60	394.99	282.00	272.50	369.00	471.00	198.50
Tot. Hardness	6.10	1640.00	143.88	152.00	74.00	120.00	174.00	100.00
NO3	0.00	42.00	1.78	4.83	0.04	0.27	1.39	1.35
HCO3	7.30	844.00	351.72	180.00	205.00	339.00	484.50	279.50
Ca	1.40	526.00	31.26	40.70	15.65	25.20	36.10	20.45
Mg	0.02	200.30	15.16	18.80	5.80	10.70	19.40	13.60
Na	1.30	1021.00	119.61	96.90	56.30	113.00	159.00	102.70
K	0.02	55.30	3.10	4.53	1.20	2.00	3.30	2.10
Cl	1.00	1509.00	48.80	121.00	7.90	15.00	42.70	34.80
SO4	0.00	3225.00	43.23	207.00	6.30	13.50	30.25	23.95
F	0.00	11.60	1.13	1.24	0.44	0.86	1.50	1.06
Temp.	29.9	34.00	30.70	0.80	30.20	30.50	31.00	0.80

[a] All units are in mg/L, except EC (µS/cm), Temp. (°C) and pH.

The concentration of fluoride in the groundwater samples ranged from 0.0 to 11.6 mg/L, with a mean value of 1.13 mg/L and a standard deviation of 1.24 mg/L. A relatively high percentage (23%) of the samples were found to have fluoride concentrations exceeding 1.5 mg/L, the WHO recommended maximum acceptable guideline value for drinking water. The percentage of the fluoride contaminated samples suggests that about 24,000 people in the study area that make use of wells covered by the study could be at risk of the incidence of fluorosis, if water from the affected boreholes is used for human consumption. Fifty two percent (52%) were within the range of 0.5-1.5 mg/L, recommended for good dental health and bone development. Twenty five percent (25%) of the groundwater samples were, however, found to have fluoride concentration lower than 0.5 mg/L, which makes the population prone to dental caries (Anku et al., 2009; Apambire et al., 1997). The total dissolved solids (TDS) of the groundwater samples in the area ranged from 8.0 mg/L to 3,373 mg/L, with a mean value of 395 mg/L. Ninety six percent (96%) of the groundwater samples were found to be fresh (TDS < 1000mg/L), and 4% were found to be brackish water (1000 mg/L< TDS < 10,000 mg/L). The pH of the groundwater samples ranged from 6.1 to 9.5, with a mean value of 7.8 and a standard deviation of 0.56. Based on classifications by Hounslow, 1995, two percent (2%) of the sampled boreholes were found to be moderately acidic (pH 4-6.5), forty six percent (46%) were neutral (pH 6.5-7.8), fifty one percent (51%) were moderately alkaline (pH 7.8-9), while only one percent (1%) was strongly alkaline (pH > 9). The pH of majority of the samples (90%) were however within the range of 6.5-8.5 and suitable for domestic use. Groundwater hardness in the study area also ranged from 8 mg/L to 1640 mg/L, with a mean value of 144 mg/L and standard deviation of 152 mg/L. The ground water in the study area could generally be classified as moderately hard to hard (75mg/l < hardness < 300 mg/L), as majority of the samples (71%) were in this category (Sawyer and McCarty, 1994). The groundwater temperature ranged between 30⁰C and 34⁰C, with a mean value of 31⁰C.

Results of the hydrochemical analysis show that aside the boreholes with elevated concentrations of fluoride (beyond 1.5 mg/L), groundwater in the study area based on the parameters analyzed is chemically acceptable and suitable for domestic use, given that other parameters not covered in the study are also within acceptable limits.

2.5.2 Inter-relationship between fluoride and electrical conductivity (EC), Calcium (Ca) and Magnesium (Mg)

Pearson's correlation was used to examine the relationship between fluoride concentration and electrical conductivity (EC), calcium and magnesium concentrations. A correlation coefficient of r >0.7 is considered to be strong whereas 0.3 < r < 0.7 is a moderate correlation and 0 < r < 0.3 is considered a weak correlation at a significance level (p) of < 0.05 (Adams et al., 2001).

The statistical study of the inter-relationship between fluoride concentration and electrical conductivity (EC) showed a positive relationship, however, the correlation coefficient (r=0.237) was found to be of low order. Similarly the study of the relationship between fluoride and calcium and magnesium concentrations, using data from the Saboba and Cheriponi districts where fluoride concentration in forty five percent (45%) of the groundwater samples exceeded the standard value of 1.5 mg/L, showed a negative relationship between fluoride and the two ions. The correlation coefficients were similarly found to be of low order (F vs. Ca; r=−0.123 and F vs. Mg; r=−0.288). Bivariate scatter plots between fluoride concentration and EC, magnesium and calcium are shown in Fig. 2-2 (a), (b) and (c), respectively.

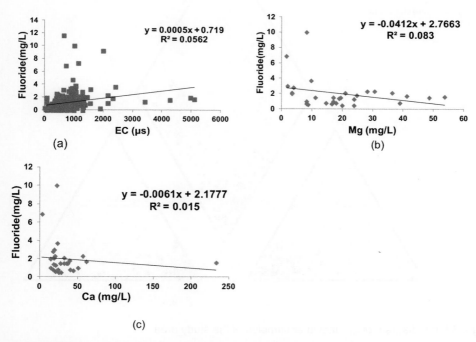

Fig 2.2 Bivariate scatter plots: (a) F vs EC, (b) F vs Mg and (c) F vs Ca.

2.5.3 Groundwater types

Using the major cations and anions, the groundwater samples were plotted on a Piper Trilinear diagram for the classification of the groundwaters in the study areas, based on their position on the diagram (Fig 2.3) (Hounslow, 1995). It was observed that the groundwater samples plotted in all the zones of the anion plot field (triangle on the right), however, the majority of the samples plotted towards the HCO_3^- corner, indicating the predominance of this anion in groundwater in the area. Similarly the samples plotted in all the zones of the cation plot field (triangle on the left), with the majority plotting towards the $Na^+ + K^+$ corner. Based on the plots on the central diamond, the water types identified in the area were: Ca-Mg-HCO₃, Ca-Mg-SO₄, Na-Cl, Na-SO₄, Na-HCO₃ and mixed water type (groundwaters in which none of the ions is dominant).

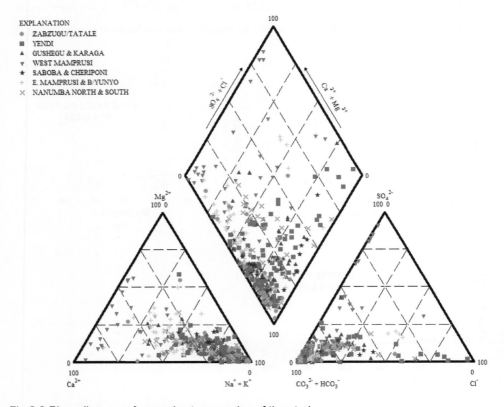

Fig 2.3 Piper diagram of groundwater samples of the study area.

2.5.4 Principal Component analysis

In this study, PCA performed on the groundwater chemical data reduced the dimensionality from the 13 original physico-chemical parameters determined in 357 groundwater samples, to 4 principal components (PCs), which cumulatively explained 72% of the data variance. Table 2.2 shows the PCs, the loadings and the percentage variance explained by each PC.

The PC loadings are used as the correlation between the original physico-chemical variables and the principal components (Vlassopoulos, et al., 2009; Praus, 2007). In hydrochemical applications, the PCs can be interpreted in terms of geochemical processes such water-rock interactions, by an examination of the loadings of the original chemical parameters on each of the PCs (Vlassopoulos et al., 2009; Chen et al., 2007). The first PC which has the highest Eigen value and accounts for the highest variance usually represents the most important process or mix of processes controlling the hydrochemistry (Yidana et al., 2010).

Looking at the loadings within the individual components (Table 2.2), it is observed that the first component (PC 1) shows strong positive relationships with Mg^{2+}, Ca^{2+}, SO4 and total hardness. The dominance of these parameters on PC1 may be a reflection of the dissolution of gypsum ($CaSO_4.2H_2O$), anhydrite ($CaSO_4$) and dolomite ($CaMg(CO_3)_2$), found in the sediments in the study area (with evaporate deposites). The second component (PC2), is mainly dominated by high positive loading for EC, TDS, Na^+, and Cl-. PC2 most likely represents a combination of processes, including the dissolution of halite (NaCl) found in the evaporates, and high rates of evapotranspiration typical of arid climates, which concentrates groundwater constituents, resulting in increased TDS.

Table 2.2 Principal components, loadings and percentage variance explained.

Groundwater chemical parameter	Principal components			
	PC1	PC2	PC3	PC4
pH	-.235	.123	.378	.615
EC	.420	.859	.151	.068
TDS	.388	.873	.144	.023
Total hardness	.937	.109	.060	-.073
NO3	.065	- .096	-.620	.212
HCO3	.144	.092	.756	.117
Ca	.736	.212	-.001	-.093
Mg	.900	-.020	.120	.015
Na	.205	.828	.300	.163
K	.108	.007	-.339	.782
Cl	-.040	.865	-.317	-.057
SO4	.862	.144	.016	.071
F	-.225	.453	.237	-.012
% Variance explained	27	25	12	8
Cumulative % of variance explained	27	52	64	72

PC3 is also characterized by positive loadings of Na^+, HCO_3^- and pH, and a negative loading for NO_3^-. The chemical parameters dominating this component suggest dissolution reactions involving carbonate and silicate minerals, cation exchange processes involving Na-montmorillonite, as well as evapotranspiration processes. The dissolution of carbonate minerals such as calcite and dolomite may result in the predominance of calcium, magnesium and bi-carbonate ions particularly in recharge areas. The replacement of calcium and magnesium ions in the water by sodium ions from the aquifer matrix through cation exchange processes with increasing residence time, may result in increased sodium ion concentration in the groundwater (Reddy, 2010; Mamatha, 2010; Guo et al., 2007; Adams et al., 2001; Freeze and Cherry, 1979). The dissolution of sodic plagioclase feldspars which occurs in the Voltain sediments, in acidic environments generated from CO_2^- charged waters, may also be a mechanism contributing to the sodium ions in the groundwater (Reddy, 2010; Hounslow, 1995; Freeze and Cherry, 1979; Acheampong and Hess, 1998). The positive loading of pH on PC 3 may be a reflection of the consumption of protons in the mineral dissolution reactions, which may result in a rise in pH. The increased pH could also be contributed from evapotranspiration processes. In natural groundwater, dissolved CO_2 combines with the water to form carbonic acid which tends to enhance the hydrogen ion concentration. During evapotranspiration processes however, some CO_2 is lost from the groundwater system alongside water, which may result in a rise in pH (Freeze and Cherry, 1979; Handa, 1975). The negative loading for NO_3^- possibly indicates an absence of contamination of the groundwater from anthropogenic activities.

The fourth component (PC4) is mainly influenced by K and pH. The predominant processes represented by PC4, may be a reflection of the importance of weathering reactions involving silicate minerals, possibly K-feldspar which is reported as occurring in the Voltain sediments (Acheampong and Hess, 1998), as well as degassing of CO_2 from the groundwater due to evapotranspiration processes.

Bivariate scatter plots for PC scores calculated for the groundwater samples for PC1, PC2 and PC3 which together account for 63% of the data variance are shown in Fig 2.4 (a), (b) and (c). The values of the scores indicate the correlations between the individual groundwater samples and the PCs (Deveral, 1989). The PC score for a sample can be related to the intensity of the geochemical processes represented by that PC, and can be interpreted as a measure of the influence of that process on the groundwater chemistry at the sample site. Extreme negative scores (<-1) reflect samples essentially unaffected by the process, and positive scores (>+1) reflect samples most affected by the process. Near-zero scores approximate samples from areas affected to an average degree by the geochemical process represented by the particular PC (Yammani et al., 2008; Senthilkumar et al., 2008). It is observed from Fig 2.4 that the processes represented by PC1 and PC2 show higher intensities at some sample sites, and wider variations in intensities in the study area compared to processes represented by PC3. The chemistry of majority of the samples however appear to be influenced by similar intensities of the three process, since they cluster around the same area as observed in Fig 2.4.

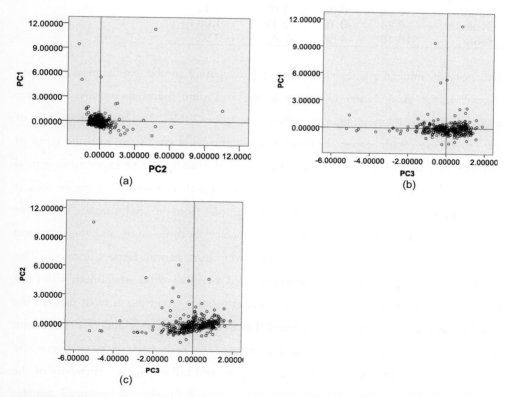

Fig.2.4 Bivariate plots of the different PCs: (a) PC1 vs PC2, (b) PC1 vs PC3, and (c) PC2 vs PC3.

2.5.5 Saturation indices of possible reactant minerals and equilibrium state of groundwater

The descriptive statistics of saturation indices calculated for all the groundwater samples using the PHREEQC software are presented in Table 2.3, which indicates the means and ranges of the indices for various mineral phases.

Table 2.3 Saturation indices (SI) for mineral phases.

| Mineral | SI | | | |
	Min.	Max.	Mean	Standard deviation
Anhydrite	-8.44	-0.38	-3.32	1.09
Aragonite	-3.27	1.91	0.12	0.73
Calcite	-3.13	2.05	0.25	0.73
Dolomite	-6.16	3.00	0.54	1.50
Fluorite	-6.39	0.95	-1.90	1.11
Gypsum	-8.24	-0.18	-3.12	1.09
Halite	-10.16	-4.68	-7.42	0.82

The possible mineral phases that may be participating in the geochemical processes controlling the groundwater chemistry in the area were found to include: anhydrite, aragonite, calcite, dolomite, fluorite, gypsum and halite (Table 2.3). The saturation state of the groundwater with respect to these minerals phases is consistent with the types of minerals that are either reported as occurring in the study area or expected in typical sedimentary formations with some evaporate deposits such as those underlying the area. It was found that about 80% of the groundwater samples from the study area are either near saturation/saturated or over-saturated with respect to both dolomite and aragonite, and about 85% were near saturation/saturated or over-saturated with respect to calcite. This suggests that the ground water has almost reached thermodynamic equilibrium with these minerals, and not much net dissolution of the minerals is occurring in most parts of the aquifers sampled. In contrast, it was found that all the samples were under-saturated with respect to halite, gypsum and anhydrite. This suggests that in the parts of the aquifers where these minerals are present, they may be dissolving and influencing the chemistry of the groundwater. Despite the relatively large percentage of fluoride-contaminated samples (about 23%) and with some high concentrations (up to 11.6 mg/L) of fluoride, about 98% of the groundwater samples were found to be under-saturated with respect to fluorite. Only

7 samples out of the 357 groundwater dataset were found to be near-saturation or over-saturated with respect to fluorite. This suggests that a net dissolution of fluorite may be occurring in the groundwater system in the areas where the mineral is present. The under saturation with respect to fluorite in some of the samples may also be due to either the low availability or absence of fluorine-bearing minerals in some locations in the aquifer system.

The results of the PHREEQC calculations and saturation states of the groundwater samples are consistent with, and supports some of the suggested geochemical processes that may be controlling the groundwater chemistry in the study area as identified from the PCA.

2.5.6 Genesis of high and low fluoride groundwater in the eastern corridor of Northern Region of Ghana

The presence of fluoride at concentrations up to 11.6 mg/L in the groundwater samples suggests that favorable conditions exist for the dissolution of fluorine-bearing minerals that may be present in some parts of the study area, as well as other processes which results in fluoride enrichment. A review of the PCA results, the equilibrium state of the groundwater based on the saturation indices, and the direct relationship between fluoride and calcium concentrations (Fig 2.2) highlight the following with regards to fluoride:

(i) fluoride has positive loadings on PC2 and PC3 which are dominated by Na, Cl, HCO_3 and pH,

(ii) fluoride shows a negative relationship on PC1 which is characterized by high positive loadings for Ca and Mg ions,

(iii) fluoride concentration is negatively related to calcium concentration,

(iv) fluoride show an inverse relationship with NO_3 as indicated by the opposite loadings of these two chemical parameters on PC3, and

(v) almost all the groundwater samples are under-saturated with respect to fluorite, and a large percentage are saturated with respect to calcite (as found from an evaluation of the saturation indices).

Fig. 2.5 is a plot of the saturation indices for fluorite and calcite for all the groundwater samples collected in the area.

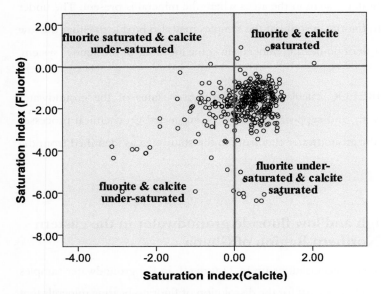

Fig.2.5 A plot of saturation indices for calcite and fluorite.

A consideration of these observations and the underlying geology/mineralogy, help to explain the genesis of high fluoride groundwaters in some parts of the eastern corridor of the Northern region of Ghana as well as the low fluoride content in some areas as discussed below.

2.5.7 High fluoride groundwaters

According to Frencken et al. (1992) and Mamatha (2010), sedimentary rocks including sandstones, shales, clays, limestone and evaporates such as those underlying the study area are typically known to contain fluoride in the range of 50-1000 mg/Kg. The fluorine is present as fluorite (CaF_2) in the carbonate rocks and evaporates, and may also be found in micas (Mamatha, 2010; Brunt et al., 2004). Both calcite and fluorite contain calcium, hence their solubilities are interdependent. As a result, conditions or processes that result in calcite precipitation and the removal of calcium from solution, promote the dissolution of fluorite and the enrichment of fluoride in solution (Mamatha, 2010; Rafique, 2009, Appelo and

Potmas, 2005). The saturation of a majority of the groundwater samples with respect to calcite and its possible precipitation, and the largely under-saturation of the samples with respect to fluorite, as shown in Fig 2.5, suggest that dissolution of fluorite may be occurring in parts of the area where it is present, resulting in fluoride enrichment.

The positive loading of fluoride for PC2 and PC3 suggests that the geochemical processes contributing to the presence of the parameters associated with PC2 and PC3 may also be the processes influencing the enrichment of fluoride in some parts of the study area. These include mineral dissolution reactions, ion exchange processes and evapotranspiration processes. These processes which result in the predominance of bicarbonate ions and increase in pH, possibly influence the attainment of the saturation limit of calcite and therefore its precipitation, a condition which favors the dissolution of fluorite and fluoride enrichment. Moreover the Na^+/Ca^{2+} cation exchange processes that may be contributing to the Na^+ concentrations associated with PC2 and PC3, also depletes the Ca^{2+} concentration in the groundwater, which promotes the dissolution of fluorite as a possible mechanism for F^- mobilization in the area (Mamatha, 2010; Appelo and Potmas, 2005; Brunt et al., 2004; Weinstein & Davison, 2004). The positive loading of fluoride on PC3 also suggests the release of fluoride from mineral surfaces into the groundwater under high pH conditions (high OH^- content) through F^-/OH^- anion exchange processes. Because of the similarity of their ionic radii (F^- = 1.23–1.36Å, OH^- =1.37–1.40Å) and geochemical behaviors, hydroxyl ions can substitute for fluoride ions on mineral surfaces.

In addition to the interactions between groundwater and fluorine-rich minerals, evapotranspiration may also be another important factor contributing to the occurrence of high fluoride groundwater in the study area. The evapotranspiration process contributes to an increase in the concentration of calcite ($CaCO_3$) as a consequence of the removal of water from groundwater in the process. This may facilitate the attainment of its saturation limit and subsequent precipitation, promoting the dissolution of fluorite and fluoride enrichment of the groundwater. Moreover, the evapotranspiration process may also directly concentrate the fluoride ions through a similar process for concentrating calcite, resulting in elevated fluoride concentrations. There is probably no contribution from anthropogenic activities to the fluoride enrichment in the study area, considering the inverse relationship between fluoride and NO_3^-, as indicated by the opposite loadings of these two chemical parameters on PC3.

2.5.8 Low fluoride groundwaters in parts of the study area

The negative relationship between fluoride and calcium concentrations (Fig.2.2) suggests that the aqueous concentrations of the two chemical parameters in some parts of the study area may be controlled by fluorite solubility and hence may obey the solubility product:

$$K_{fluorite} = [Ca^{2+}][F^-]^2 = 10^{-10.57} \qquad\qquad (2.1)$$

at equilibrium, where $[Ca^{2+}]$ and $[F^-]$ denote calcium and fluoride molar concentrations respectively and $K_{fluorite}$, represents fluorite solubility product.

Consequently in areas where the calcium ion concentration is high, it may play a critical role in determining the fluoride content of the groundwater as the upper limit of the fluoride concentration may be determined by the Ca ion concentration in accordance with equation (2.1). That is the dissolution of fluorite may be suppressed when the concentration of Ca is above the limit for fluorite solubility, resulting in low fluoride content (Apambire et al. 1997; Appelo and Potmas, 2005; Chae et al, 2007). It was found from the PCA that the loading of fluoride for PC1 was negative (Table 3), which suggest that the geochemical processes or reactions associated with that component (PC1) contribute to the depletion of fluoride from the groundwater in the areas where the reactions are occurring, resulting in low fluoride content. The geochemical processes which include the dissolution of gypsum and anhydrite, result in the predominance of Ca ions in the groundwater, a favorable condition which may either suppress the dissolution of fluorite or cause its precipitation. This may explain the low content of fluoride in some parts of the groundwater system. The low fluoride content or absence in some of the groundwater samples may also be due to the low availability or absence of fluorine-bearing minerals in some locations (Kundu et al., 2001).

2.5.9 Relationship of fluoride to geology and the spatial distribution

Results of the relationship between the underlying geology and fluoride distribution which was examined using the ArcView spatial join procedure is shown in Table 2.4.

Table 2.4: Fluoride occurrence related to geology.

Geological unit		No. of boreholes	Fluoride (mg/L)			
Formation	Dominant lithology		Min.	Max.	Mean	Standard deviation
Upper Voltain	Sandstone, conglomerate, thin beds of shale and mudstone.	40	0.1	0.8	0.3	0.18
Middle Voltain (Obossom and Oti beds)	Mudstone, shale, sandstone, conglomerate, limestone and arkose.	277	0	11.6	1.27	1.3
Lower Voltain	Quartz sandstone and grits.	1	0.02	0.02	0.02	0
Buem	Shale, sandstone, lava, and tuff with limestone, grit and conglomerate	7	0.1	1.6	0.61	0.5

The spatial distribution of fluoride in the ground water samples from the study area is displayed in Fig. 2.6.

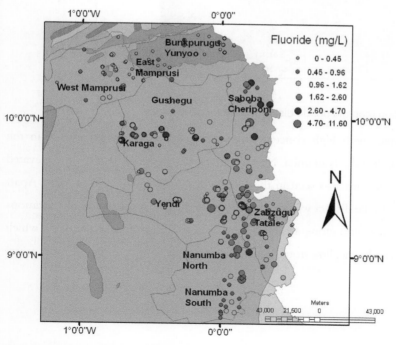

Fig. 2.6 Spatial distribution of fluoride in the study area.

Fig. 2.7 shows the Inverse distance weighted (IDW) prediction map of fluoride distribution in the study area, which may be a useful guide for determining the fluoride levels in the non-sampled/measured parts of the study area.

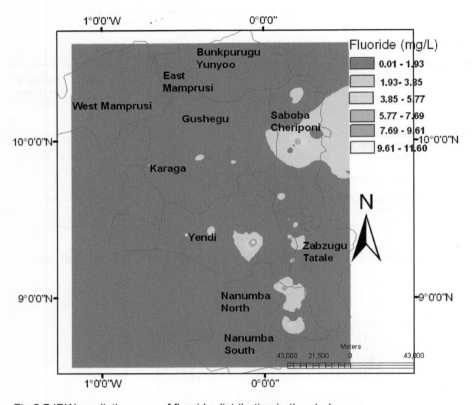

Fig.2.7 IDW prediction map of fluoride distribution in the study area.

Groundwater samples with high concentrations of fluoride are located mostly in the Saboba, Cheriponi, Yendi, Nanumba North and South districts. Patches of elevated fluoride concentrations are also seen in the Karaga and Zabzugu/Tatale districts. Apart from the Zabzugu/Tatale district which is underlain by the Buem formation, the locations with high fluoride concentrations are underlain by the Obossom and Oti beds which consists mainly of sandstone, limestone, conglomerate, shale, arkose and mudstone.

2.5.10 Variation of fluoride with depth

Pearson's correlation was also used to examine a possible relationship between fluoride concentration and groundwater depth in the study area.

The statistical study showed a weak but significant positive correlation ($r = 0.179$) between the two parameters (i.e. fluoride concentration and groundwater depth). Fig 2.8 shows a plot between fluoride concentration and depth of groundwater samples.

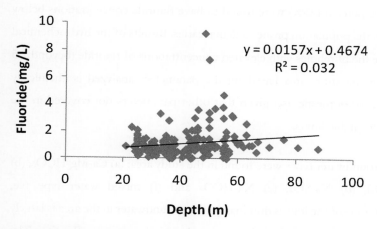

Fig. 2.8 A plot of fluoride concentration versus groundwater depth.

The positive correlation between fluoride concentration and borehole depth suggests that geochemically more favorable conditions for fluoride enrichment possibly prevail at deeper depths in the aquifers of the area. According to Karro et al.(2006), the pH of groundwater and the contents of Na^+ and Cl^- generally increase with depth and the groundwater changes towards the Na-Cl-HCO_3 chemical type, which provides favorable conditions for fluoride enrichment. The increase of residence time with increasing depth may also increase the contact time of the groundwater with fluorine-bearing minerals, which enhances their dissolution and release of fluoride (Chae et al, 2007).

2.6 Conclusions

Twenty three percent (23%) of 357 groundwater samples from the eastern corridor of the Northern region of Ghana, for which the physico-chemical parameters were studied were found to have fluoride concentrations exceeding 1.5 mg/L, the WHO guideline for drinking water, with concentrations as high as 11.6 mg/L. Human consumption of water from these wells can result in the incidence of fluorosis. Fifty two percent (52%) of the groundwater sample were within the acceptable fluoride concentration range of 0.5-1.5 mg/L, while twenty five percent (25%) were found to have fluoride concentrations below 0.5 mg/L, which makes the population prone to dental caries. Results of the hydrochemical analysis show that aside the boreholes with elevated concentrations of fluoride (beyond 1.5 mg/L), groundwater in the study area based on the parameters analyzed is chemically acceptable and suitable for domestic use, given that other parameters not covered in the study are also within acceptable limits.

The following major groundwater types were found in the study area: (a) Ca-Mg-HCO$_3$, (b) Ca-Mg-SO$_4$, (c) Na-Cl, (d) Na-SO$_4$, (e) Na-HCO$_3$ and (f) mixed water type (i.e, groundwaters in which none of the ions is dominant). The groundwater in the area is largely saturated with respect to calcite and under-saturated with respect to fluorite. The geochemical processes that control the overall groundwater chemistry in the study area possibly also promote fluoride enrichment of the groundwater. These processes include mineral dissolution reactions, ion exchange processes and evapotranspiration. The predominant mechanisms controlling the fluoride enrichment probably include; calcite precipitation and Na/Ca exchange processes both of which depletes Ca from the groundwater and promotes the dissolution of fluorite. The mechanisms also include F$^-$/OH$^-$ anion exchange processes, as well as evapotranspiration processes which concentrates the fluoride ions and increases its concentration. Elevated fluoride levels in the study area were found to occur predominantly in the Saboba and Cheriponi districts and also in the Yendi, Nanumba North and South districts. These areas are underlain by the Obossom and Oti beds comprising mainly of sandstone, limestone, conglomerate, shale, arkose and mudstone. The results of the current study may help in the planning of strategies for the provision of safe drinking water in both cases of excessive and insufficient fluoride content in the groundwater in the study area.

References

Acheampong, S, Y., Hess, J.W. 1998. Hydrogeological and hydrochemical framework of shallow groundwater system in the southern Voltain Sedimentary Basin, Ghana. Hydrogeology J., 6(4), 527-537.

Adams, S., Titus, R., Pietersen, K., Tredoux, G., Harris, C. 2001. Hydrochemical characteristics of aquifers near Sutherland in the Western Karoo, South Africa. J. of hydrology, 241, 91-103.

Anani, C. 1999. Sandstone petrology and the provenance of the Neoproterozoic Voltain Group in the southern Voltain Basin, Ghana. Sedimentary Geology, 128, 83-89.

Anku, Y.S., Banoeng-Yakubo, B., Asiedu, D.K., Yidana, S.M. 2009. Water quality analysis of groundwater in crystalline basement rocks, Northern Ghana. Envrin Geol., 58, 989-997.

Anongura, R.S. 1995. Fluorosis survey of Bongo district of upper east region- Ghana: Report submitted to Action research committee, University of Development Studies, Tamale.

Apambire, W. B., Boyle, D. R., Michel, F. A. 1997. Geochemistry, genesis, and health implications of fluoriferous groundwaters in the upper regions of Ghana. Env. Geol., 33 (1), 13-24.

Appelo, C.A.J and Postma, D. 2005. Geochemistry, groundwater and pollution. 2nd edition, CRC Press, New York.

Asiamah, R.D., Senayali, J.K., Agjei-Gyapong, T and Spaargaren, O.C. 1997. Ethno-pedology surveys in the semi-arid Savanazone of Northern Ghana. An ILEIA initiated project report 97/04. Soil Research Institute, Kwadaso-Kumasi and International Soil Reference and Information Centre (ISRIC), Wageningen.

Ayoob, S., Gupta, A.K. 2009. Performance evaluation of alumina cement granules in removing fluoride from natural synthetic waters. *Chemical Engineering Journal*, 150: 485-491

Ayugane, C. 2008. The effect of high fluoride levels in rural water supplies on inhabitants in six selected communities in the Saboba and Cheriponi districts of the Northern region: A report to the Community Water and Sanitation Agency of Northern region (CWSA-NR), Tamale, Ghana.(Unpublished).

Biswas, K., Bandhoyapadhyay, D., Ghosh, U.C. 2007. Adsorption kinetics of fluoride on iron (III) zirconium hybrid oxide. Adsorption, 13, 83-94.

British Goelogical Survey, 2009. Water quality fact sheet: Fluoride, British Geological Survey.

Brunt, R., Vasak, L., Griffion, J. 2004. Fluoride in groundwater: Probability of occurrence of excessive concentration on global scale. igrac International groundwater Resources Assessment Centre.

Cattel, R.B. 1966. The scree test for number of factors. Multivariate Behav. Res., 1,275-276

Chae, G., Yun, S., Mayer, B., Kim, K., Kim, S., Kwon, J., Kim, K., Koh, Y. 2007. Fluorine geochemistry in bedrock groundwater of South Korea. Science of the Total Environment, 385, 272-283.

Chen, K., Jiao, J.J., Huang, J., Huang, R. 2007. Multivariate statistical evaluation of trace elements in groundwater in a coastal area in Shenzhen, China. Environmental Pollution, 147, 771-780.

Chen, N., Zhang, Z., Feng, C., Zhu, D., Yang, Y., Sugiura, N. 2010. Preparation and characterization of granular ceramic containing dispersed aluminum and iron oxides as adsorbents for fluoride removal from aqueous   solution. Journal of Hazardous Materials, 186, 863-868

CWSA-NR. 2007. Summary of the fluoride issues in the northern region: Report submitted to the Government of Ghana. Community Water and Sanitation Agency, June 2007, Tamale, Northern region (unpublished).

Deveral, S.J. 1989. Geostatistical and principal component analysis of ground chemistry and soil salinity data, San Joaquin Valley, California. Regional characterization of water quality, Proceedingd of Baltimore Symposium, May 1989. IAHS.

Fawell, J., Bailey, K., Chilton, J., Dahi, E., Fewtrell, L., Magara, Y. 2006. Fluoride in drinking water. IWA Publishing, London. 4-81.

Freeze, R., Cherry, J. 1979. Groundwater. Prentice Hall, New Jersey.

Frencken, J.E. (editor).1992. Endemic Fluorosis in developing countries, causes, effects and possible solutions. Publication number 91.082, NIPG-TNO, Leiden, The Netherlands.

Gao, X., Wang, Y., Li, Y. 2007. Enrichment of fluoride in groundwater under the impact of saline water intrusion at the salt lake area of Yuncheng basin, Northern China. Env. Geol, 53, 795-803.

Gill, H.E. 1969. A groundwater reconnaissance of the Republic of Ghana, with description of geohydrolic provinces: US Geological Survey Water Supply Paper.

Guo, Q; Wang, Y; Ma, T; Ma, R. 2007. Geochemical processes controlling the elevated fluoride concentrations in groundwaters of the Taiyuan Basin, Northern China. Journal of Geochemical Exploration, 93, 1-12.

Gyau-Boakye, P and Dapaah-Siakwan, S. 2000. Groundwater as a source of rural water supply in Ghana. J. of Applied Sci. and Technol., 5(1&2), 77-86.

Handa, B.K. 1975. Geochemistry and genesis of fluoride-containing groundwaters in India. Groundwater, 13(3), 275-281.

Harder, R. 2008 Fluoride-Toxin or medicine. Department of Water Management, Delft University of Technology, Delft, Netherlands.

Hounslow, A.W (1995) Water quality data: Analysis and interpretation. CRC Press Inc. Floriida, U.S.

Hurtado, R., Gardea-Torresdey,J., Teiman, K.J. 2000. Fluoride occurence in tap water at Lod altos de Jalisco in the Central Mexico region. Proceedings of the 200 Conference on Hazardous Waste Research.

Junner, N.R., Hirst, T. 1946. The geology and hydrogeology of the Volta Basin. Gold Coast Geological Survey, Memoir 8

Junner, N. R., Service, H. 1936. Geological notes on Volta River District and Togoland under British mandate. Annual Report on the Geological Survey by the Director, 1935–1936

Kaiser, H.F. 1960. The application of electronic computers to factor analysis. Educational and Physiological Measuement, 20 (1), 141-151.

Karro, E., Indermitte, E., Saava, A., Haamer, K., Marandi, A. 2006. Fluoride occurrence in publicly supplied drinking water in Estonia. Environmental Geol. DOI 10.1007/s00254-006-0217-1.

Kesse, G.O. 1985. The Mineral and Rock Resources of Ghana. Balkema, Rotterdam.

Kundu, N., Panigrahi, M.K., Tripathy, S., Munshi, S., Powel, M.A., Hart, B.R. 2001. Geochemical appraisal of fluoride contamination of groundwater in the Nayagarh distract of Orissa, India. Environmental Geology, 41, 451-460.

Madhnure, P., Sirsikar, D.Y., Tiwari, A.N., Ranjan, B., Malpe, D.B. 2007. Occurrence of fluoride in the groundwaters of Pandharkawada area, Yavatmal district, Maharashtra, India. Current Science, 92(5):675-678.

Mamatha, P., Rao, S.M. 2010. Geochemistry of fluoride rich groundwater in Kolar and Tumkur districts of Karnataka. Environ Earth Sci, 61, 131-142.

Mandinic, Z., Curcic, M., Antonijevic, B., Carevic, M., Mandic, J., Djukic-Cosic, D., Lekic, C.P. 2010. Fluoride in drinking water and dental fluorosis. Sci. of the Total Env. , 408, 3507-3512

Mjengera, H.J., Mkongo, G.B. 2009. Occurrence of fluoride in water sources and water defluoridation in Tanzania. Ministry of Water, Dar es Salaam, Tanzania.

Naani, A., Roisenberg, A., Fachel, J.M.G., Mesquita, G., Danieli, C. 2008. Fluoride characterization by principal component analysis in the hydrochemical facies of Serra Geral Aquifer System in Southern Brazil. Annals of the Brazelian Academy of Science, 80(4), 693-701.

Negendra C R (2003) Fluoride and Environment—A Review. Third international conferenceonenvironment and health, Chennai, India, 15-17 December, 2003.

Parhurst, D.L and Appelo, C.A.J. 1999. PHREEQC for Windows version 1.4.07. A hydrochemical transport model. U.S. Geological Survey Software.

Praus, P. 2007. Urban water quality evaluation using multivariate analysis. Acta Montanistica Slovaca, 12 (2), 150-158.

Rafique, T., Naseem, S., Usmani, T.H., Bashir, E., Khan, F.A, Bhanger, M.I. 2009. Geochemical factors controlling the occurrence of high fluoride water in the Nagar Parkar area, Sindh, Pakistan. Journal of Hazardous Materials, 171, 424-430

Reddy, D.V., Nagabhushanam, P., Sukhija, B.S., Reddy, A.G.S., Smedley, P.L. 2010. Fluoride dynamics in the granitic aquifer of the Wailapally watershed, Nalgonda District, India. Chem. Geol., 269, 278-289.

Rogers, R.J. 1989. Geochemical comparison of groundwater in areas of New England, New York and Pennsylvania. Groundwater, 27(5), 690-712.

Senthikumar, G., Ramanathan, A.L., Nainwal, H.C., Chidambaram, S. 2008. Evaluation of the hydrogeochemistry of groundwater using factor analysis in the Cuddalore coastal region, TamilNadu, India. Indian Journal of Marine Science, 37(2), 181-185.

Shomar, B., Muller, G., Yahya, A., Aska, S., Sansur, R. 2004. Fluoride in groundwater, soil and infused black tea and the occurrence of dental fluorosis among school children of the Gaza strip. J. of Water and Health, 2 (1): 23-35.

Tiemann M(2006) Fluoride in drinking water: a review of fluoridation and regulation issues.http://www.ncseonline.org/NLE/CRSreports/o6Mar/RL33280.pdf. Cited September 2009.

Vlassopoulos, V., Gion, J., Zeliff, M., Porcello, J., Tolan, T., Lindsey, K. 2009. Groundwater geochemistry of the Columbia River Basalt group aquifer system: Columbia basin groundwater management area of Adams, Franklin, Grant, and Lincon Counties, Washinton, USA.

Walther, I .2009. Fluoride contamination in the main Ethiopian Rift Valley. http://www.geo.tu-friberg.de/fernerkundung/PAPERS/fluor_walther.pdf. Cited September, 2009

Weinstein, L.H, Davison, A (2004) Fluoride in the environment: Effects on plants and animals. CABI Publishing, Wallingford, U.K.

WRC. 2006. Hydrogeological Assessment of Northern regions of Ghana: A bibliographic review of selected papers. Water Resources Commission, Accra, Ghana.

Yammani, S.R., Reddy, T, V, K., Reddy, M.R.K. 2008. Identification of influencing factors for groundwater quality variation using multivariate analysis. Env. Geol., 55, 9-16.

Yidana, S.K. 2010. Groundwater classification using multivariate statistical methods: Southern Ghana. J. of African Earth Sci., 57, 455-469.

Yidana, S.K., Ophori, D., Banoeng-Yakubo, B. 2008. Hydrochemical evaluation of Voltain system-The Afram Plains area, Ghana. J. of Env. Manag., 88, 697-707.

Yidana, S.M; Banoeng-Yakubo, B; Akabzaa, T.M. 2010. Analysis of groundwater quality using multivariate and spatial analysis in the Keta basin, Ghana. J. of African Earth Sci., 58, 220-21234.

3

Drinking water defluoridation using aluminium (hydr) oxide coated pumice: Synthesis, equilibrium, kinetics and mechanism

Main part of this chapter was published as:

Salifu, A., Petrusevski, B., Ghebremichael, K., Modestus, L., Buamah, R., Aubry, C., Amy, G.L. 2013. Aluminum (hydr)oxide coated pumice for fluoride removal from drinking water: Synthesis, equilibrium, kinetics and mechanism. Chem. Eng. J. 228, 63 -74.

Chapter 3 Drinking water defluoridation using aluminium (hydr) oxide
coated pumice: Synthesis, equilibrium, kinetics and mechanism

69

Abstract

Modification of pumice particle surfaces by aluminum oxide coating was found effective in creating hard surface sites for fluoride adsorption, in accordance with the hard and soft acids and bases (HSAB) concept. Aluminum oxide coated pumice (AOCP) reduced fluoride concentration in model water from 5.0 ±0.2 mg/L to 1.5 mg/L in approximately 1 hr, in batch adsorption experiments using an adsorbent dose of 10 mg/L. Contrary to expectations, thermal treatment of AOCP aimed at further improving its performance, instead reduced the fluoride removal efficiency. The equilibrium adsorption of fluoride by AOCP conformed reasonably to five isotherm models in the order: Generalized model > Langmuir type 2 > BET >Temkin >Dubinin-Radushkevich; with a Langmuir maximum adsorption capacity of 7.87 mg/g. AOCP exhibited good fluoride adsorption within the pH range, 6-9, which makes it possible to avoid pH adjustment with the associated cost and operational difficulties, especially if it is to be used in remote areas of developing countries. Based on results from kinetic adsorption experiments, it was observed that at a neutral pH of 7.0 ±0.1 which is a more suitable condition for groundwater treatment, fluoride adsorption by AOCP was fairly faster in the initial period of contact than a grade of activated alumina (AA) that was tested, the commonly used adsorbent for water defluoridation. AOCP is thus promising and could also possibly be a useful fluoride adsorbent.

3.1 Background

There are two properties of chemical species, i.e the electronic chemical potential (μ) and chemical hardness (η), that determine their chemical behavior, and which leads to broadly useful principles. In this theory, μ and η for chemical species (a collection of nuclei and electrons such as ions, molecules, radicals) are defined respectively as the first and second derivatives of the electronic energy of the ground state (E) with respect to the number of electrons (N) at constant external potential, given by equations (3.1) and (3.2):

$$\mu = \left(\frac{\delta E}{\delta N}\right)_v \tag{3.1}$$

$$\eta = \left(\frac{\delta^2 E}{\delta N^2}\right) = \frac{1}{2}\left(\frac{\delta \mu}{\delta N}\right)_v \tag{3.2}$$

where v is the potential due to nuclei attraction and any other external forces.

In practice, however, the method of finite difference is used to determine approximations for μ and η, which relates these parameters to the ionization energy (I) and electron affinity (A) of the chemical species by equations (3.3) and (3.4):

$$-\mu \approx \frac{(1+A)}{2} = X \tag{3.3}$$

$$\eta \approx \frac{(1-A)}{2} \tag{3.4}$$

where χ is the absolute electronegativity, the negative of the chemical potential ($-\mu$) (Pearson, 1993; Person 1988; Makov, 1995; Vigneresse et al., 2011; Vigneresse, 2012; Alfarra et al., 2004).

Reactions between chemical species occur through electron transfer. Consider two given reactant species, an acid (A) and a base (B), the flow of electrons will be from the species of lower absolute electronegativity,χ (the base) to that of higher χ (the acid), until the electronic chemical potential (μ) becomes equal at equilibrium. For reactants A and B, their

Chapter 3 Drinking water defluoridation using aluminium (hydr) oxide
coated pumice: Synthesis, equilibrium, kinetics and mechanism

71

chemical potentials μ_A and μ_B after electron transfer between them can be presented by equations (3.5) and (3.6):

$$\mu_A = \mu_A^\circ + \eta_A \Delta N \tag{3.5}$$

$$\mu_B = \mu_B^\circ + \eta_B \Delta N \tag{3.6}$$

where, μ_A° and μ_B° are their neutral state chemical potentials and ΔN is the number of electrons transferred. At equilibrium for the acid - base system;

$$\mu_A = \mu_B \tag{3.7},$$

therefore;

$$\Delta N = \frac{(\mu_B^0 - \mu_A^0)}{(\eta_A + \eta_B)} = \frac{(\chi_A^0 - \chi_B^0)}{(\eta_A + \eta_B)} \tag{3.8}$$

Equation (3.8) indicates that, the differences in absolute electronegativities drives the electron transfer process (or reaction), and the sum of hardness acts as resistance to the process. Thus the chemical hardness of a system is a measure of the resistance to changes in its electronic configuration and together with the electronegativity is thought to be an indicator of chemical reactivity and stability (Pearson, 1993; Person 1988; Makov, 1995; Vigneresse, 2011; Vigneresse, 2012; Alfarra et al., 2004).

Equation (3.8) also gives insight into the hard-soft acid-basis (HSAB) concept, which is one of the fundamental constructs in chemistry. According to the HSAB concept, due to their characteristics, hard acids prefer binding to hard bases to give predominantly ionic complexes, whereas soft acids prefer binding to soft bases to form predominantly covalent complexes. From equation (3.8), a soft acid and a soft base would have a large number of electron transfer (ΔN), which are used to form a covalent bond between them. In contrast, a hard acid and a hard base would have a small ΔN and tend to form ionic bonding instead. The soft-soft interactions with covalent character and hard-hard interaction with ionogenic character results in the most stable interactions and those reactions are therefore more preferred (Pearson, 1995; Vigneresse et al., 2011; Makov, 1995).

Due to its characteristic, and with a hardness value of 7.0 ev, a ranking order of hardness values determined for bases, shows F- as the hardest base and would therefore prefer to coordinate to a hard acid. A similar ranking of hardness values for hard acids also show B^{3+} ($\eta = 110.72$), Be^{2+} ($\eta = 67.84$), Al^{3+} ($\eta = 45.77$) as being the hardest of the acids, and would have strong affinity towards F- in accordance with the HSAB concept (Alfra et al., 2004; Vigneress, 2009; Pearson, 1993). Al^{3+} is considered to be readily available, presumably in most developing countries, due to the use of aluminum bearing salts such as $Al_2(SO_4)_3$ in conventional water treatment. Thus with regards to the search for appropriate materials for drinking water defluoridation, Al^{3+} could be a suitable and sustainable source of hard acid for binding of fluoride ions and their subsequent removal from aqueous environments. This could contribute to prevention of the incidence of fluorosis and other fluoride-related health hazards in most developing countries where there is presence of excess fluoride in the groundwater used for drinking and cooking. Al^{3+} does not occur naturally in the elemental form, but readily combines with oxygen and water to form oxides and/or hydroxides (Rosenqviast, 2002).

The aim of the present work was therefore to modify the surfaces of pumice particles by aluminum oxide coating in order to create hard surface sites for fluoride adsorption from drinking water in accordance with the HSAB concept.

The suitability of pumice as base/support material for the surface modification process was studied, since use of pumice could contribute to cost reductions and sustainable solutions for water deflouridation, especially in developing countries where it is indigenous. Pumice is a volcanic material with a rough surface and porous structure which may provide large number of attachment sites for aluminum coating, and therefore is expected to result in good fluoride removal efficiency and high adsorption efficiency.

The fluoride adsorption potential of the aluminum oxide coated pumice (AOCP) was investigated in batch adsorption experiments. The effect of thermal treatment of AOCP on

Chapter 3 Drinking water defluoridation using aluminium (hydr) oxide
coated pumice: Synthesis, equilibrium, kinetics and mechanism

73

its fluoride removal efficiency was also investigated. The effects of adsorbent dose and pH on fluoride removal by AOCP were examined. The physico-chemical characteristics of AOCP were determined. Additionally initial results from this study, i.e., the fluoride adsorption kinetics of AOCP, were compared with that of activated alumina (AA) which is commonly used for water defluoridtaion (Ghorai and Pant, 2005; Reardon and Wang, 2000).

3.2 Materials and Methods

3.2.1 Synthesis of Al-oxide coated pumice

3.2.1.1 Coating of aluminum oxides onto pumice

Pumice samples used in this work as base material for the aluminum oxide coating were obtained from Aqua-TECHNIEK bv, The Netherlands. The samples were sieved to a particle size range of 0.8-1.12mm, a common size range used in drinking water filtration, thoroughly washed with demineralized water, and air-dried before being used in the coating process. To coat the pumice, a sufficient amount of 0.5 M $Al_2(SO_4)_3$ was added to completely soak about 150 g of the dried pumice in a beaker. The mixture was stirred mechanically for 1.5 hours at 150 rpm and pumice was after that drained and air dried at room temperature. The dried pumice was subsequently soaked in 3M NH_4OH to neutralize it and complete the coating process. The coated sample was sieved again and washed several times with demineralized water buffered at pH 7.0± 0.1, in order to remove any loosely bound aluminum oxides. AOCP was finally dried and stored for fluoride adsorption studies.

3.2.1.2 Thermal treatment of AOCP

Samples of AOCP were calcined at temperatures between 200⁰C and 1000⁰C, in a muffle furnace for 2 h. The thermal treatment was aimed at further enhancing the fluoride removal capacity (Kawasaki et al., 2008).

3.2.2 Characterization techniques

The aluminum contents of both uncoated pumice and AOCP samples were extracted by acid digestion at a temperature of 200°C using reagent grade concentrated HNO_3 acid, and was measured by colorimetric methods using a UV-VIS Recording spectrophotometer (SHIMADZU, Japan).

The specific surface area of both uncoated pumice and AOCP were determined by N_2 gas adsorption-desorption method at 77 K with TriStar 3000 gas adsorption analyzer (Micromeritics, USA), and the Brunauer-Emmett-Teller (BET) method was used in the calculation. The Barrett-Joyner-Halenda (BJH) pore size model was used for determining pore size distribution. The empirical t-plot methodology was used to discriminate between contributions from microspores and remaining porosity (i.e mesoporosity, macroporosity and external surface area contributions). Prior to the N_2 adsorption measurements, the samples were pre-treated by degassing in a vacuum at $1,500^0C$ (Groen et al., 2003).

Chemical compositions were obtained from X-Ray Fluorescence (XRF) analysis provided by Axiosmax-Advance (from PANalytical Spectris, U.K). Spot elemental analysis of AOCP was carried out using a scanning electron microscope (SEM NovaNano from FEI company, Netherlands) coupled with an energy dispersive X-ray spectroscopy (EDX) (EDAX, AMETEX, USA). Point of zero charge (pHpzc) of AOCP was determined by a mass titration method (Bourika et al., 2003).

3.2.3 Fluoride adsorption experiments

A 1000 mg/L stock fluoride solution was prepared by dissolving reagent grade NaF in demineralized water. Model fluoride water used for adsorption experiments was prepared by diluting the stock solution with Delft (The Netherlands) tap water with the composition of major cations and anions (mg/L) as presented in Table 3.1, to obtain a fluoride concentration of 5 ± 0.2 mg/L, similar to excess fluoride found in some groundwater wells in Northern region of Ghana. The bicarbonate concentration was adjusted to 260 mg/L by spiking with sodium bicarbonate in order to simulate similar concentrations in groundwater in the Northern region of Ghana.

Chapter 3 Drinking water defluoridation using aluminium (hydr) oxide coated pumice: Synthesis, equilibrium, kinetics and mechanism

75

Table 3.1 Composition of Delft tap water (mg/L).

Parameter	Ca	Mg	SO_4^{2-}	Cl	Na	K
Conc.(mg/L)	49.7	9.0	78.8	72.2	35	5.5

Batch adsorption experiments were conducted using the completely mixed bath reactor method for studying the fluoride removal behavior of both AOCP and activated alumina (AA). Kinetic experiments were performed using a constant AOCP or AA dose of 10 g/L and aqueous samples were taken at pre-determined times. Isotherm experiments were conducted for AOCP, employing the variable adsorbent dose method (1 g/L to 20 g/L), which has been suggested to provide more reliable isotherm data for determining adsorption capacity, compared to variable concentration method (Ayoob and Gupta, 2008)

The adsorption experiments were conducted at a neutral pH 7.0 ± 0.1 and room temperature (20°C), using 0.5 L of model water in 500 ml P.E bottles. HCl and/or NaOH were used for pH adjustment after which the P.E bottles were tightly closed to avoid exchange of CO_2 in order to stabilize pH during the experiments. The top covers of the P.E. bottles were designed to hold a syringe for sampling, such that contact with CO_2 was avoided during the sampling. A control adsorption experiment using uncoated pumice was conducted. In addition, a blank fluoride adsorption experiment, without adsorbent, was carried out to evaluate experimental uncertainties. All adsorption experiments were carried out in duplicate in order to check reproducibility. The experiments were carried out on a shaker (Innova 2100 Platform shaker, New Brunswick Scientific, USA), at a shaking speed of 100 rpm. Aqueous samples taken at pre-determined times during the adsorption experiments were immediately filtered through a microfilter paper (0.45 µm) to separate adsorbent from fluoride model water and prevent any further reactions.

The residual fluoride concentration in the filtrate was determined potentiometrically using an Ion-selective electrode (WTW F 800 DIN, Germany). The total ionic strength adjusting buffer (TISAB) was used to release any complexed fluoride ions before total fluoride measurements.

The specific amount of fluoride uptake, q_t (mg/g) at time t (h), was determined using the mass balance relationship:

$$q_t = V(C_0 - C_t)/W \tag{3.9}$$

and the fluoride removal efficiency (% adsorption) was calculated using the equation:

$$\% \text{ adsorption} = 100 \times (C_o - C_t)/C_o \tag{3.10}$$

where C_0 and C_t (mg/L) are initial fluoride concentration and the concentrations at time t (h), respectively, V is volume of solution (L), and W (g) is mass of adsorbent used.

3.3   Results and discussions

3. 3.1 Characterization of uncoated pumice and AOCP

The average aluminum contents of uncoated pumice and AOCP samples were 89 mg Al/g and 111 mg Al/g, respectively. The net amount of aluminum coating onto the pumice particles was thus 22 mgAl/g, indicating the effectiveness of the coating procedure.

Table 3.2 presents elemental composition of both uncoated pumice and AOCP obtained from XRF and EDX analysis. EDX spectrum of AOCP is also shown in Fig.3.1. The XRF results showed oxides of Si and Al as major constituents of both materials, while other elements were present in relatively smaller amounts. After the coating, the presence of Al was observed to increase from 21.9% to 29.4%. Emergence of sulfur in AOCP is attributed to the coating process using $Al_2(SO_4)_3$. The EDX results confirmed AOCP samples enrichment in aluminum, which is associated with appearance of sulfur and increase of oxygen abundance. This indicates that pumice has incorporated together aluminum and oxygen. Moreover, lower variations of the relative abundance in EDX compared to XRF measurements suggest that aluminum oxide was not only deposited at the surface of the grain but also penetrated inside. It is expected that the fluoride ions would be adsorbed mostly by the oxides of aluminium.

Chapter 3 Drinking water defluoridation using aluminium (hydr) oxide
coated pumice: Synthesis, equilibrium, kinetics and mechanism

77

Table 3.2: Elemental composition (% by wt.) of uncoated pumice and AOCP obtained from XRF and EDX characterizations.

Elements		C	O	Na	Al	Si	K	Fe	S	Ca
EDS (wt%)	Uncoated pumice	7.5	42.1	7.1	11.1	25.8	4.6	1.8	-	-
	AOCP	3.1	44.8	6.1	13.5	24.2	4.3	2.1	0.9	0.5
XRF (wt%)	Uncoated pumice	-	-	12.8	21.9	46.3	8.4	3.7	0.1	1.6
	AOCP	-	-	10.7	29.4	40.2	8.1	3.8	2.5	1.3

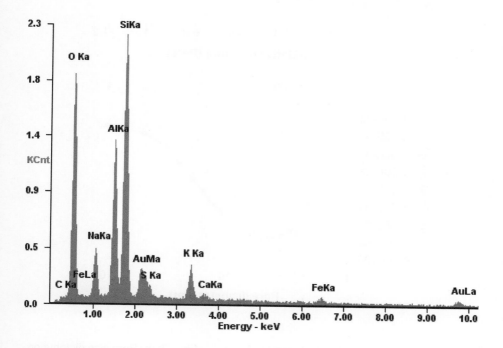

Fig.3.1: EDX spectrum of AOCP.

Fig.3.2 (a) shows the N_2 adsorption-desorption isotherm of AOCP, which is classified as a type 1V isotherm according to the IUPAC classification (Kaneko, 1994), and characterizes the presence of mesopores. The pore-size distribution (Fig 3.2(b)) also showed that the pore sizes of AOCP mostly varied between 5 and 100nm, which further revealed AOCP as a typical mesoporous- macroporous material by the IUPAC classification. The radius of fluoride is 1.33 Å, which is much smaller than the pore size range (5 -100nm), and, it is therefore conducive for the penetration of fluoride ions into the inner layers of AOCP.

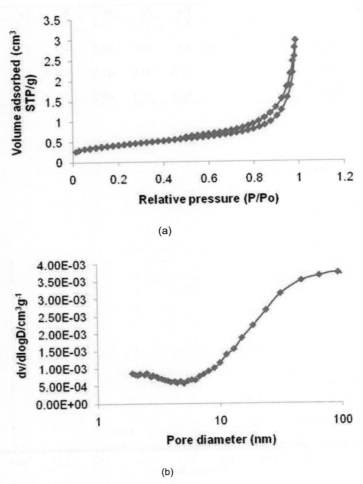

(a)

(b)

Fig.3. 2: (a) N_2 adsorption-desorption isotherm for AOCP.   Fig.3.2: (b) Pore size distribution for AOCP.

The weight loss upon pre-treatment, BET surface area (S_{BET}), total pore volume (V_{pore}), micropore volume (V_{micro}), and external surface area (mesopore area) (S_{meso}) are shown in Table 3.3.

Chapter 3 Drinking water defluoridation using aluminium (hydr) oxide
coated pumice: Synthesis, equilibrium, kinetics and mechanism

79

Table 3.3 Surface area and porosity of AOCP and uncoated pumice.

Pumice sample	Weight loss (wt %)	S_{BET} (m²/g)	V_{pore} (cm³/g)	V_{micro} (cm³/g)	S_{meso}
Uncoated pumice	0.4	3.4	-	-	-
AOCP	6.3	1.5	0.005	<0.001	1.0

It was observed that the aluminum coating decreased the surface area of pumice by 55%. The decrease in surface area was possibly due to a significant degree of micropores blockage by the Al coating.

The pHpzc of AOCP determined by mass titration was 5.2.

3.3.2 Efficacy of AOCP for water defluoridation

From initial studies conducted to examine its effectiveness, AOCP was found able to reduce fluoride concentration in model water from 5.0 ± 0.2 mg/L to 1.5 mg/L in approximately 1 h (Fig 3.3), using AOCP dose of 10 g/L. Uncoated pumice was not effective for fluoride removal even though it also contained aluminum (Fig 3.3). The efficacy of AOCP for fluoride adsorption is likely due to modifications of pumice particle surfaces by the aluminum coating. Presumably, the aluminum content of uncoated pumice is inert.

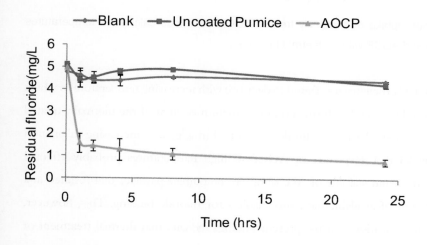

Fig. 3.3 Fluoride removal by uncoated pumice and AOCP in batch adsorption experiment: Model water: fluoride = 5 ± 0.2 mg/L, HCO₃ = 260 mg/L, pH =7.0±0.1, adsorbent dose = 10 g/L, shaker speed = 100 rpm

3.3.3 Fluoride adsorption efficiency of thermally treated AOCP

Contrary to expectations, thermal treatment of AOCP samples aimed at further improving the fluoride removal performance (Kawasaki et al., 2008) was instead found to reduce the removal capability, and resulted in lower fluoride removal than that of AOCP without thermal treatment (Fig. 3.4(a)).

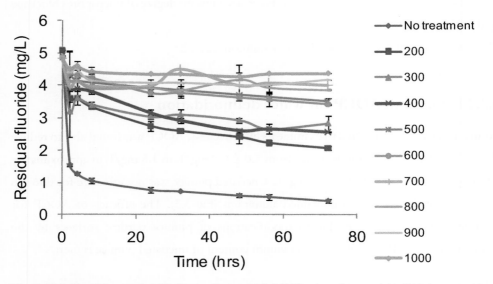

Fig. 3.4(a) Fluoride uptake by thermally treated AOCP at various calcination temperatures (200°C -1000°C) and AOCP with no thermal treatment.

The fluoride removal efficiency was found to decrease with increasing temperature to which AOCP was exposed (Fig.3.4(b)). Extraction of aluminum content of the thermally treated AOCP also showed a similar trend with decrease of Al (mg/g) with increasing calcination temperature (Fig.3.4 (b)). Exposure of AOCP to high temperatures probably affected negatively, the nature and stability of Al coating on the pumice particles that resulted in a reduction in amount of available hard surface sites for fluoride binding. This, however, requires further investigation, but the present findings indicate that thermal treatment of AOCP in its synthesis process may not augment for fluoride removal.

Chapter 3 Drinking water defluoridation using aluminium (hydr) oxide coated pumice: Synthesis, equilibrium, kinetics and mechanism

81

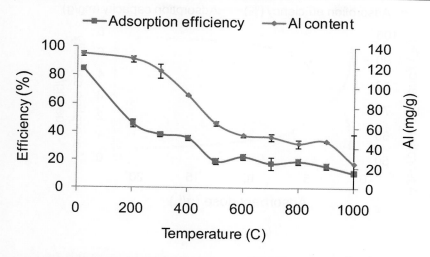

Fig. 3.4 (b) Effect of thermal treatment on fluoride removal efficiency of AOCP and aluminum content.

3.3.4 Effect of AOCP dose

The effect of AOCP dose on the fluoride removal was studied using different masses of adsorbent (1 - 20 g/L) and a fixed fluoride concentration of 9.5 ± 0.2 mg/L. The results are presented in Fig 3.5 as fluoride (a) adsorption efficiency (%) and capacity, q_e (mg/g) as functions of adsorbent dose, and (b) the distribution coefficient (K_d) as function of adsorbent dose, respectively.

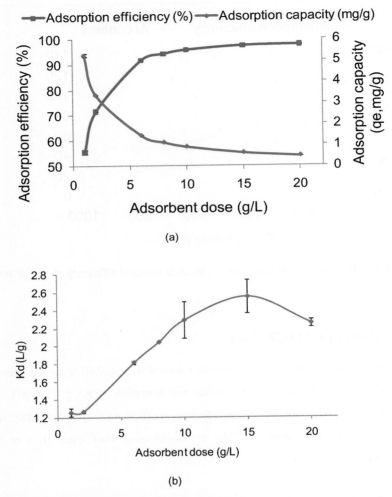

(a)

(b)

Fig.3.5: (a) Fluoride adsorption efficiency (%) and adsorption capacity, q_e (mg/g) as functions of adsorbent dose, and (b) K_d as function of adsorbent dose.

It was found that the fluoride removal efficiency increased from 55.7% to 97.8% when the adsorbent dose was increased from 1 to 20 g/L (Fig 5a). This is likely due to the greater availability of active surface sites for F- binding as a consequence of increased adsorbent dose. The increase of removal efficiency was however found to be negligible after an AOCP dose of 10 g/L, which may be considered the optimum. On the other hand the adsorption capacity, q_e, (mg/g) of AOCP decreased from 5.3 to 0.5 mg/g with increased adsorbent dose for the fixed fluoride concentration (Fig. 5a). This observation agrees well with an

Chapter 3 Drinking water defluoridation using aluminium (hydr) oxide
coated pumice: Synthesis, equilibrium, kinetics and mechanism

83

increasing solid dose for a fixed solute load as this may result in the availability of fewer fluoride ions per unit mass of AOCP, i.e, lesser fluoride/AOCP ratio with increasing adsorbent dose (Wang et al., 2009). The decrease of adsorption capacity with increasing adsorbent dose also suggests heterogeneity of AOCP surface sites. According to the surface site heterogeneity model, the adsorbent surface is composed of sites with a wide spectrum of binding energies. At low dosage for such adsorbents, all types of sites are entirely exposed for fluoride adsorption and the surface get saturated faster resulting in a higher q_e value. At higher adsorbent dose, however, the availability of higher energy sites decreases with a large fraction of lower energy sites occupied which results in a lower q_e (Biswas et al., 2009).

The distribution coefficient, K_d reflects the binding ability of an adsorbent surface for a solute. K_d is a ratio of solute concentration in the solid phase and in aqueous phase, and the values for a given system depend mainly on solution pH and type of adsorbent surface. The K_d values for fluoride and AOCP were calculated at pH 7.0±0.1 using the following relationship:

$$K_d = q_e/C_e \qquad\qquad (3.11)$$

where q_e (mg F /g AOCP) and C_e (mgF$^-$/L) are solute concentration in the solid adsorbent and aqueous phase at equilibrium, respectively. The values of K_d as a function of adsorbent dose are shown in Fig. 3.5b. The distribution coefficient was found to increase with an increase in adsorbent dose, which further highlights the heterogeneous nature of the AOCP surface. At a given pH, K_d values for a solute indicates two different situations with regards to the property of an adsorbent surface: (a) the K_d values increase with an increase in adsorbent dose if the surface is heterogeneous, whereas (b) the K_d values remain the same with increase in adsorbent dose if the surface is homogenous (Chen et al., 2011; Nigussie et al., 2007; Wang et al., 2009).

3.3.5 Adsorption kinetics, mechanism and rate-controlling step

Adsorption equilibrium is not an instantaneous process as the adsorbate must first diffuse from the bulk solution through various diffusion mechanisms to available active surface sites on the adsorbent for attachment (Sontheimer et al., 1998). The kinetics of adsorption describes the rate of adsorbate uptake on adsorbent and controls the equilibrium time.

In this study, fluoride adsorption kinetic experiments were performed using both AOCP and activated alumina (AA) as adsorbents. The aim was to investigate the kinetics of fluoride adsorption by AOCP, and, additionally to compare the AOCP fluoride adsorption kinetics with that of AA.

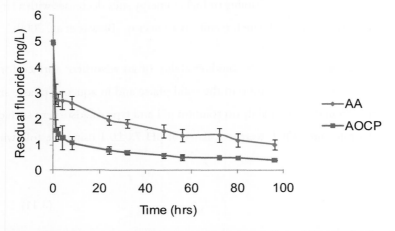

Fig 3.6: Fluoride removal by AOCP and AA as function of time: adsorbent dose = 10 g/L, AOCP and AA particle size = 0.8 - 1.12mm; model water: fluoride = 5 ± 0.2 mg/L, HCO_3 = 260 mg/L, pH = 7.0 ± 0.1

From Fig 3.6, a rapid fluoride uptake at the initial period of contact followed by a slower adsorption with time until equilibrium is observed. The initial rapid fluoride removal is presumably due to the instantaneous utilization of the most readily available active sites on the outer surface of AOCP by fluoride ions, and slower adsorption at the later stage may be due to a gradual diffusion of fluoride ions into the interior surface of the porous adsorbent for adherence. It was further observed that at a neutral pH of 7.0 ± 0.1, which is a more suitable condition for groundwater treatment particularly in remote rural areas, the kinetics of fluoride adsorption by AOCP was quite comparable to that of AA or perhaps fairly faster at the initial period of contact (Fig 3.6). AOCP reduced fluoride concentration of 5 ±0.2 mg/L to the WHO standard (1.5 mg/L) in approximately 1 h and attained more than 80% removal within 24 h.   In contrast, AA attained the WHO standard in approximately 48 h, and about 60% removal in 24 h (Fig 3.6). The fairly faster removal of fluoride by AOCP may be due to the availability of a larger number of active binding sites on the exterior surface of AOCP, compared to AA.

3.3.5.1 Adsorption kinetic modeling and rate parameters

Adsorption kinetic parameters are useful for the prediction of adsorption rate that gives important information for modeling the process and designing adsorption-based water treatment systems (Yousef et al., 2011). In the present study different kinetic models, namely, the Elovich, Largergren pseudo-first order and Ho's pseudo-second order kinetic models were used to analyze the fluoride adsorption kinetics data. The applicability of a particular model for the fluoride-adsorbent (AOCP or AA) system was evaluated from the goodness of data fit, coefficient of determination (R^2), and comparison of experimental and predicted amounts of fluoride adsorbed at equilibrium q_e (mg/g). The intra-particle diffusion model developed by (Weber & Morris, 1963) was further tested for identifying the mechanism of fluoride adsorption onto the adsorbents.

3.3.5.1.1 Pseudo-first order and second-order kinetic models

The mathematical representations (nonlinear forms) of the first and second-order kinetic models are given in equations (3.12) and (3.13), respectively, which can be integrated to obtain the respective linear forms given by equations (3.14) and (3.15), using the boundary conditions; $q_t = q_t$ at $t = t$ and $q_t = 0$ at $t = 0$.

$$q_t = (1 - e^{-k_1 t}) \qquad (3.12)$$

$$q_t = \frac{q_e^2 k_2 t}{1 + q_e k_2 t} \qquad (3.13)$$

$$\ln(q_e - q_t) = \ln q_e - k_1 t \qquad (3.14)$$

$$t/q_t = 1/h_o + 1/q_e t \qquad (3.15)$$

$$\text{where } h_o = k_2 q_e^2 \tag{3.16}$$

For equations (3.12) to (3.16), q_t and q_e represents the amount of fluoride adsorbed (mg/g) at any time t (h) and at equilibrium, respectively, $k_1 (h^{-1})$, and k_2 (g mg^{-1}h^{-1}) are the first and the second- order adsorption rate constants, respectively, and h_o is the initial adsorption rate at t $\rightarrow$ 0. The rate constant k_1 was calculated from the slope of the linear plot of In(q_e - q_t) vs. time (t) of the kinetic experimental data using equation (3.14), (Fig 3.7a), while k_2 was determined from the intercept of the plot between t/q_t against t, (Fig 3.7b), based on equation (3.15). The predicted amounts of fluoride adsorbed at equilibrium, q_e (mg/g) were also determined from the above linear plots.

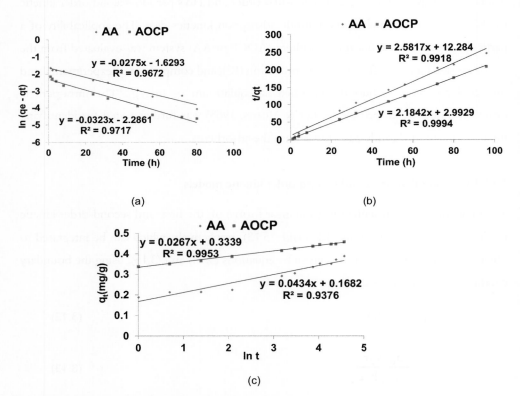

Fig. 3.7: Adsorption kinetics of fluoride onto AA and AOCP: (a) First-order plot, (b) Second-order plot and (c) Elovich.

Chapter 3 Drinking water defluoridation using aluminium (hydr) oxide coated pumice: Synthesis, equilibrium, kinetics and mechanism

87

The rate constant k_2 was then used to calculate the initial adsorption rate (h_o) according to equation (3.16) (Maliyekkai et al., 2006; Fan et al., 2003; Dawood and Sen, 2012). The rate parameters for first and second-order kinetic models, the corresponding R^2 values, the experimental and predicted q_e values as well as the initial adsorption rate (h_o) are presented in Table 3.4.

The adsorption kinetics data for both AA and AOCP fitted better with the pseudo-second-order kinetic model based on the R^2 values (Table 3.4). Moreover for both adsorbents, the pseudo-first-order kinetic model predicted significantly lower values of equilibrium adsorption capacity $q_{e,1(cal.)}$ than the experimental values $q_{e(exp.)}$, which suggests that the adsorption process of fluoride onto these adsorbents is not a first-order reaction. On the other hand the theoretical $q_{e,2(cal.)}$ values determined from the second-order model were in good agreement with the experimental values ($q_{e(exp.)}$) (Table 3.4), which further shows the applicability of the model to describe the adsorption process of fluoride onto both AA and AOCP (Febrianto et al., 2009). The model is based on the assumption that the fluoride adsorption process follows a second-order chemisorption.

Table 3.4 Kinetic parameters for fluoride adsorption by AOCP and AA

Adsorbent	First-order-kinetic model			Second-order-kinetic model			q_e (mg/g) (exp.)	h_o (mg/g h)	$t_{1/2}$ (hr)
	k_1 (1/h)	q_e(mg/g) (cal.)	R^2	k_2 (g/mg/h)	q_e(mg/g) (cal.)	R^2			
AA	0.028	0.196	0.9672	0.543	0.387	0.9918	0.388	0.081	4.76
AOCP	0.032	0.102	0.9717	1.594	0.458	0.9994	0.458	0.334	1.37

Table 3.4 (continued)

Adsorbent	Elovich				Intra-particle diffusion rate constant, K_d (mg/g h$^{1/2}$)	C (mg/g)	R^2	D_P (cm²/h x10^{-5})	D_F (cm²/h x10^{-7})
	α	β	$1/\beta\ln(\alpha\beta)$	R^2					
AA	2.092	23.041	0.168	0.9376	0.022	0.173	0.9859	5.81	3.69
AOCP	8709.94	37.453	0.334	0.9953	0.013	0.3406	0.9495	20.2	15.34

Chapter 3 Drinking water defluoridation using aluminium (hydr) oxide
coated pumice: Synthesis, equilibrium, kinetics and mechanism

89

Half-adsorption time, $t_{1/2}$ which is defined as the time required for the adsorbent to take up half as much fluoride as its equilibrium value, is often used as a measure of the adsorption rate (Dawood & Sen, 2012). For the second-order kinetic process, $t_{1/2}$ can be calculated from equation (3.17):

$$t_{1/2} = 1/k_2q_{e,2} \qquad\qquad (3.17)$$

The values of $t_{1/2}$ calculated for both AA and AOCP are shown on Table 3.4.

3.4.5.1.2 Elovich model

The Elovich model (equation (3.18)), is one of the useful kinetic models for describing chemisorption, where α and β are constants during a specific experiment. The constant α (mg/g min.) is considered as the initial rate because $\frac{dq_t}{dt}$ approaches α as q_t approaches zero. Integration of equation (3.18) with the same boundary conditions as for the pseudo-first and second-order kinetic models results in the linearized form, equation (3.19).

$$\frac{dq_t}{dt} = \alpha e^{-\beta q_t} \qquad\qquad (3.18)$$

$$q_t = 1/\beta \ln(\alpha\beta) + 1/\beta \ln t \qquad\qquad (3.19).$$

A plot between the adsorbed fluoride q_t against time (t) based on equation (3.19) should yield a straight line, (if applicable), with slope $(1/\beta)$ and intercept $1/\beta \ln(\alpha\beta)$. The β value is indicative of the number of sites available for adsorption and $1/\beta \ln(\alpha\beta)$ is the adsorption quantity when $\ln t = 0$, i.e when $t = 1h$. Thus the $1/\beta \ln(\alpha\beta)$ value helps in understanding the adsorption behavior in the initial period (Jimenez-Nuznez et al., 2012; Sivasankar et al., 2011; Sivasankar et al., 2012; Tseng & Tseng, 2006; Ozacar & Sengil, 2005; Tseng, 2006).

It was found that the kinetics data for the AA/AOCP - F- adsorption systems were well described by the Elovich model (Fig 3.7c), which further supports the involvement of chemisorption in the adsorption mechanism. The Elovich parameters, including the adsorption quantity in the initial period, $1/\beta \ln(\alpha\beta)$ are presented in Table 3.4. The $1/\beta$ $\ln(\alpha\beta)$ value for AOCP was observed to be fairly higher than that of AA.

3.3.5.1.3 Intra-particle diffusivity analysis and rate-controlling step

Understanding the underlying mechanism of fluoride adsorption and rate controlling step is important for the determination of an appropriate residence time, which is essential for good process design and control of adsorption treatment systems (Yousef et al., 2011).

Theoretically the mechanism of fluoride removal from aqueous phase by porous adsorbents such as AA and AOCP is assumed to consist of four essential steps, namely, (i) migration of fluoride ions from bulk solution to the boundary layer or film surrounding the adsorbent particle (i.e bulk transport), (ii) diffusion of fluoride ions across the boundary layer to exterior surface of the adsorbent particles, called external or film diffusion, (iii) transport of adsorbate from the exterior surface to the interior surface of adsorbent via surface and pore diffusion mechanisms, termed intra-particle diffusion and (iv) uptake (adsorption) of fluoride ions on available sites either on the exterior or interior surface of adsorbent. The slowest step determines the overall fluoride uptake rate and is the rate controlling step (Yousef et al., 2011; Dawood & Sen, 2012; Faust & Aly, 1998). The adsorption step (step iv) is usually rapid, hence does not influence the overall kinetics. Moreover for an adsorbate-solution system that is sufficiently agitated during the interaction period, the rate is not limited by bulk diffusion (step I). Thus the overall rate may be controlled by either film and/or intra-particle diffusion (Faust & Aly, 1998). The technique most commonly used for identifying the mechanism involved is by fitting the experimental data into an intra-particle diffusion model, a kinetic-based model proposed by (Weber and Morris, 1963), which is represented by the following relationship:

$$q_t = K_{id}t^{1/2} + C \tag{3.20}$$

where K_{id} (mg/gh$^{1/2}$) is the intra-particle diffusion rate constant and C (mg/g) is a constant that is proportional to the boundary layer thickness. The larger the C value, the greater the boundary layer thickness (Singh et al., 2012). If intra-particle diffusion is involved in the

Chapter 3 Drinking water defluoridation using aluminium (hydr) oxide
coated pumice: Synthesis, equilibrium, kinetics and mechanism

91

adsorption process, a plot of q_t vs. $t^{1/2}$ will be linear, and if the line passes through the origin, then the rate-limiting process is only due to intra-particle diffusion. Otherwise some other mechanism is also involved in the rate-controlling step alongside intra-particle diffusion (Yousef et al., 2011).

Fig 3.8 illustrates the plots of kinetic experimental data for fluoride adsorption onto AA and AOCP based on equation (3.20), and the intra-particle diffusion rate constants as well as the C values are presented in Table 3.4.

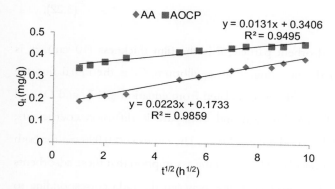

Fig. 3.8 Intra-particle diffusion plots for adsorption of fluoride onto AA and AOCP.

It is noted from Table 3.4 that intra-particle diffusion plays a significant role in the adsorption of fluoride onto both AA and AOCP based on the R^2 values. The linear plot did not however pass through the origin for both AA and AOCP (Fig 3.8), which indicates the process of fluoride adsorption onto these adsorbents is a complex one, and both film and intra-particle diffusion contributes to the rate-determining step. The adsorption rate is film diffusion controlled during the initial stages of the adsorption process, in which the fluoride ions diffuse across the boundary layer surrounding the adsorbent particles, and are adsorbed at the outer surface of the adsorbent. As the adsorbent outer surface become loaded with fluoride ions, the process in the later stages become controlled by intra-particle diffusion through the pores for uptake at the interior surface of the adsorbent until equilibrium is attained. The boundary layer thickness, C, also provides insight into the tendency for the fluoride ions to either adsorb onto the adsorbent phase or remain in solution. Higher values of C depict higher adsorption. In this study, the C value for AOCP was observed to be fairly higher than that of AA (Table 3.4).

The diffusion coefficient, D, whose values largely depend on the surface properties of the adsorbent, is an important parameter indicative of the diffusion mobility. The diffusion coefficients (Table 3.4) for the film diffusion (D_F), and for intra-particle transport (D_P) of fluoride ions within the pores of AA and AOCP particles were calculated using equations (3.21) and (3.22) as follows:

$$D_F = 0.23r_o \partial q_e/(t_{1/2}C_o) \tag{3.21}$$

$$D_P = 0.03r_o^2/t_{1/2} \tag{3.22}$$

where r_o (cm) is the radius of adsorbent particle, ∂ is the film thickness (10^{-3}cm), q_e is concentration of fluoride in solid phase (mg/g) at equilibrium, C_o is the initial fluoride concentration (mg/L), $t_{1/2}$ is the half-life as calculated from equation (3.17), and D_F and D_P are the values (cm^2/h) of film diffusion and intra-particle diffusion coefficients, respectively (Biswas et al., 2009; Dawood & Sen, 2012). The radius $r_o = 0.096$ cm for both adsorbents (AA and AOCP) was calculated based on the assumption that these adsorbents consist of spherical particles with an average radius between the radii corresponding to upper- and lower-size fractions (0.08 - 0.112 cm). The film diffusion coefficient (D_F) and intra-particle diffusion coefficient (D_P) values are shown in Table 3.4.

A comparison of the various kinetic parameters as presented on Table 3.4, give indications that fluoride adsorption onto AOCP was fairly faster than that of AA.

3.3.6 Equilibrium isotherms for fluoride removal by AOCP

The isotherm deals with the relationship between equilibrium amounts of fluoride on adsorbent (AOCP) and concentration in solution. Analysis of isotherm data is important for determining adsorption capacity of adsorbents, one of the main parameters required for designing an adsorption treatment system (Sakar et al., 2006). Inherent assumptions in the development of isotherm models reported in literature, however, generally limits their application under diverse conditions. Thus it becomes helpful to compare different

Chapter 3 Drinking water defluoridation using aluminium (hydr) oxide
coated pumice: Synthesis, equilibrium, kinetics and mechanism

93

isotherm models in the process of selecting a best-fitting model for a given experimental data set (Ho, 2004; Ayoob & Gupta, 2008).

To model the fluoride adsorption onto AOCP, the equilibrium experimental data were fitted to five isotherm models; namely Langmuir, Dubinin-Radushkevich, Temkin, Generalized and BET equations, in order to find the model that best fit experimental data. Table 3.5 presents the isotherm models and their corresponding linearized forms (Kumar & Sivanesan, 2006; Foo & Hamid, 2012; Malakootian et al., 2011; Karadag et al., 2007; Allen et al., 2003; Liu & Liu, 2008; Nemr, 2009; Ozmihci & Kargi 2006). The coefficients of determination (R^2) values, were used to determine the best-fit isotherm model (Kumar & Sivanesan, 2006; Allen et al., 2003).

Table 3.5 Isotherms and their linear forms.

Isotherm	Nonlinear form	Linear form	Plot	Parameters/Constants
Langmuir Type 1	$q_e = \dfrac{q_{max}bC_e}{1+bC_e}$	$\dfrac{C_e}{q_e} = \dfrac{1}{q_{max}}C_e + \dfrac{1}{bq_{max}}$	$\dfrac{C_e}{q_e}$ vs. C_e	$q_{max} = \dfrac{1}{slope}$; $b = \dfrac{slope}{intercept}$
Langmuir Type 2		$\dfrac{1}{q_e} = \left(\dfrac{1}{bq_{max}}\right)\dfrac{1}{C_e} + \dfrac{1}{q_{max}}$	$\dfrac{1}{q_e}$ vs. $\dfrac{1}{C_e}$	$q_{max} = \dfrac{1}{intercept}$; $b = \dfrac{intercept}{slope}$
Langmuir Type 3		$q_e = q_{max} - \left(\dfrac{1}{b}\right)\dfrac{q_e}{C_e}$	q_e vs. $\dfrac{q_e}{C_e}$	$q_{max} = intercept$; $b = \dfrac{-1}{slope}$
Langmuir Type 4		$\dfrac{q_e}{C_e} = bq_m - bq_e$	$\dfrac{q_e}{C_e}$ vs. q_e	$q_{max} = \dfrac{-intercept}{slope}$; $b = -slope$
Temkin	$q_e = \dfrac{RT}{b}\ln(A_T C_e)$	$q_e = B_T\ln A_T + B_T\ln C_e$; $B_T = \dfrac{RT}{b}$	q_e vs. $\ln C_e$	$B_T = slope$; $b_T = \dfrac{RT}{slope}$; $\ln A_T = \dfrac{intercept}{slope}$
Dubinin-Radushkevich	$q_e = (q_s)\exp(-K_{DR}\varepsilon^2)$; $\varepsilon = RT\ln\left(1+\dfrac{1}{C_e}\right)$	$\ln q_e = \ln q_s - K_{DR}\varepsilon^2$	$\ln q_e$ vs. ε^2	$K_{DR} = slope$; $\ln q_s = intercept$
BET	$q_e = \dfrac{q_s K_B C_e}{(C_s - C_e)\left[1+(K_B-1)\left(\dfrac{C_e}{C_s}\right)\right]}$	$\dfrac{C_e}{(C_s - C_e)q_e} = \dfrac{1}{K_B q_s} + \left(\dfrac{K_B-1}{K_B q_s}\right)\left(\dfrac{C_e}{C_s}\right)$	$\dfrac{C_e}{q_e(C_s - C_e)}$ vs. $\dfrac{C_e}{C_s}$	–
Generalized	$q_e = \dfrac{q_{max}C^{N_b}}{(K_G + C^{N_b})}$	$\log\left(\dfrac{q_{max}}{q_e} - 1\right) = \log K_G - N_b\log C_e$	$\log\left(\dfrac{q_{max}}{q_e} - 1\right)$ vs. $\log C_e$	$N_b = slope$; $\log K_G = intercept$

Chapter 3 Drinking water defluoridation using aluminium (hydr) oxide coated pumice: Synthesis, equilibrium, kinetics and mechanism

95

3.3.6.1 Langmuir isotherm

The basic assumption in deriving the Langmuir model is that adsorption occurs at specific homogenous sites within the adsorbent by monolayer sorption, without interaction between adsorbate molecules. It further assumes that once an adsorbate occupies a particular site, no further adsorption can occur at that site. Thus in theory, a saturation point is reached where all active sites are occupied, and no further adsorption can occur. Monomolecular layer formation assumed in the derivation of the Langmuir isotherm model is associated with chemisorption process. As indicated in Table 3.5, the Langmuir isotherm model can be linearized into four different types, and simple linear regression could result in different estimates of isotherm parameters for the same experimental data, depending on which of the linearized forms is fitted to the data (Foo & Hamid, 2010; Malakootian et al., 2011; Karadag et al., 2007). This is because each of the different linearization methods of the non-linear isotherm model (i.e original Langmuir isotherm expression) to the corresponding linear form changes the original error distribution, either for better or worse, which can impact the estimation of isotherm parameters (Kinniburg, 1986; Kumar, 2006a; Kumar, 2006b). Among the different linearized Langmuir equations, type 1 and 2 are the most commonly used by many researchers for fitting adsorption equilibrium data due to the minimized deviations from the fitted curve, resulting in the best error distribution) (Kumar & Sivanesan, 2006). The Langmuir isotherm constants are q_{max} (mg/g), the maximum adsorption capacity which corresponds to a complete monolayer coverage on adsorbent (AOCP), and b (L/mg), a constant related to the affinity of the binding sites. For each of the four linearized isotherm equation, the isotherm constants (q_{max} and b) can be determined from the slope and intercept, of the plot based on the respective equation as indicated on Table 3.5.

For the Langmuir model, the shape or characteristics of the isotherm can be used to predict the favorability of the adsorption process under given experimental conditions. The essential characteristics of the isotherm can be determined by a dimensionless constant separation factor, R_L, defined by equation (3.23) (Ghorai & Pant, 2005):

$$R_L = 1/(1+bC_o) \tag{3.23}$$

where b is the Langmuir constant, and C_o is the initial fluoride concentration. A value of $R_L < 1$ represents favorable adsorption and $R_L > 1$ represents unfavorable adsorption (Chen et al., 2011; Ghorai & Pant; 2005; Malakotian, 2011).

Table 3.6 shows the isotherm parameters, R_L and R^2 values obtained for the four linearized Langmuir equations for fluoride removal by AOCP. Based on the R^2 values, the experimental data were found to be described best by the Langmuir type 2 equation (highest R^2 value) with $q_{max} = 7.87$mg/g, while the other three types (1, 3 & 4) showed a very poor fit (low R^2) to the experimental data. Langmuir type 3 and 4 equations showed similar R^2 values, which suggests that the two linear forms have similar error distribution structure (Kumar & Sivanesan, 2006). The R_L value obtained indicate that fluoride adsorption on AOCP is favorable for the Langmuir isotherm under the experimental conditions.

Chapter 3 Drinking water defluoridation using aluminium (hydr) oxide coated pumice: Synthesis, equilibrium, kinetics and mechanism

97

Table 3.6 Summary of parameters obtained from different isotherms for fluoride adsorption by AOCP.

Isotherm model/parameter	Isotherm parameter value
Langmuir type 1:	
q_{max} (mg/g)	11.82
b (L/mg)	0.057
R_L	0.781
R^2	0.531
Langmuir type 2:	
q_{max} (mg/g)	7.874
b (L/mg)	0.087
R_L	0.696
R^2	0.985
Langmuir type 3:	
q_{max} (mg/g)	5.827
b (L/mg)	0.126
R_L	0.617
R^2	0.449
Langmuir type 4:	
q_{max} (mg/g)	11.901
b (L/mg)	0.056
R_L	0.782
R^2	0.448
Temkin	
B_T	0.860
b_T (kJ/mol)	2.640
A_T	2.271
R^2	0.956
Dubinin-Radushkevich	
K_{DR}	3×10^{-7}
q_s(mg/g)	1.652mg/g
Energy (E)(kJ/mol)	1.29
R^2	0.9131
BET	
q_s	1.5mg/g
K_B	3.867
R^2	0.9702
Generalized	
N_b	1.036
K_G	11.269
R^2	0.992

3.3.6.2 Dubinin-Radushkevich isotherm

Unlike the Langmuir equation, the Dubinin-Radushkevich (D-R) isotherm model, which is based on the Polanyi theorem, can provide information about the energy required for the adsorption process and mechanism involved. The D-R isotherm assumes a Gaussian (normal) distribution of energy sites (Abusafa & Yucel, 2002; Onyango et al., 2004). The non-linear and linearized forms of the equation are shown in Table 3.5, where q_s (mg/g) is the adsorption capacity and K_{DR} is a constant related to adsorption energy. A plot of Inq_e versus ε^2 based on the linear D-R equation should give a straight line from which K_{DR} and q_s can be determined from the slope and intercept respectively, where ε is the polanyi potential (equal to $RTIn(1+1/C_e)$, (Table 3.5) and R, T, q_e and C_e are the gas constant (8.314 j/mol K), temperature in Kelvin, equilibrium amount of fluoride on AOCP, and equilibrium concentration in solution respectively (Foo & Hameed, 2010; Malakootian et al., 2011). The mean free energy of adsorption (E) for the fluoride-AOCP system can be determined using equation (3.24), the magnitude of which may provide useful information with regards to the adsorption mechanism, whether it is a chemical or physical process (Foo & Hameed, 2010; Abusafa & Yucel, 2002).

$$E = \frac{1}{\sqrt{2K_{DR}}}. \tag{3.24}$$

Table 3.6 shows the values of q_s and K_{DR} obtained from the linear plot of the D-R equation as well as the mean free energy of adsorption (E =1.29 kJ/mol) determined from equation (3.24). Typical bonding energy range reported for ion-exchange mechanism is 8 - 16kJ/mol. It is also generally noted that values up to -20kJ/mol is indicative of physisorption process due to electrostatic interaction between charged particles, while more negative values than -40kJ/mol involve chemisorption (Zafar et al., 2007; Kirah & Kaushik, 2008). The E value (1.29 < 8 kJ/mol) obtained in this study suggests that physical adsorption plays a significant role in fluoride up-take by AOCP.

3.3.6.3 Temkin Isotherm

The derivation of the Tempkin isotherm equation is based on the

Chapter 3 Drinking water defluoridation using aluminium (hydr) oxide
coated pumice: Synthesis, equilibrium, kinetics and mechanism

99

assumptions that: (i) the heat of adsorption would decrease linearly rather than logarithmically with increase of adsorbent coverage, due to some indirect adsorbate-adsorbate interactions and (ii) adsorption is characterized by a uniform distribution of binding energies, up to a maximum binding energy. The isotherm model and its linear form are shown in Table 3.5, where A_T (L/g) is equilibrium binding constant related to maximum binding energy, $B_T = RT/b_T$ and b_T (kJ/mol) is Temkin constant related to heat of adsorption, while R and T are as defined for the D-R equation. The constant b_T reflects the bonding energy which in turn indicate the type of interaction between fluoride and AOCP. The Temkin isotherm constants,

which were obtained from the slope and intercept of a plot of q_e versus $\ln C_e$, using the linear equation (Yousef et al., 2011; Foo & Hameed, 2010; Malakootian et al., 2011; Allen et al., 2003; Liu & Liu, 2008) are presented on Table 3.6. The order of magnitude of the energy value ($b_T = 2.271$ kJ/mol) obtained from the Temkin isotherm also suggests the involvement of physisorption process in the interaction between fluoride and AOCP, which is similar to the predictions from the D-R isotherm.

3.3.6.4 Brunauer-Emmett-Teller (BET) isotherm

The BET isotherm model (Table 3.5) is a theoretical equation developed for multimolecular adsorption, which is an extension of the Langmuir model, where K_B is the equilibrium constant which characterizes the interaction energy of adsorbate with the surface of adsorbent, q_s is maximum monolayer capacity (mg/g) and C_s is adsorbate monolayer saturation concentration (mg/L) (Febriato et al., 2009; Foo & Hameed, 2010; Malakootian et al., 2011; Karadag, 2007; Liu & Liu, 2008; Kiran & Kaushik, 2008). The isotherm constants obtained from a plot of $\dfrac{C_e}{q_e(C_s-C_e)}$ vs $\dfrac{C_e}{C_s}$ based on the linearized BET equation, are presented in Table 3. 6.

3.3.6.5 Generalized isotherm model

Another model which was tested for describing the equilibrium data for fluoride adsorption onto AOCP was the generalized adsorption equation (Table 3.5).The isotherm parameters: K_G(mg/L), the saturation constant and N_b, the cooperative binding constant (Table

3.6),were determined from the intercept and slope respectively of a plot of $\log\left(\frac{q_{max}}{q_e} - 1\right)$ vs. $\log C_e$, in accordance with the linearized form of the equation. The q_{max} used in the generalized equation was as obtained from the Langmuir type 2 isotherm and $q_e(mg/g)$ and C_e (mg/L) are equilibrium concentration in the solid and liquid phases, respectively (Malakootian et al., 2011; Nemr, 2009; Ozmihci and Kargi, 2006).

It was observed that the AOCP-fluoride adsorption equilibrium data conformed reasonably well to the five isotherm models in the order, based on R^2 values: Generalized model > Langmuir type 2 > BET >Temkin >Dubinin-Radushkevich; with a maximum adsorption capacity of 7.87 mg/g (Table 3.6). Results from the isotherm studies indicated physical adsorption plays a significant role in the adsorption mechanism, while the kinetic studies indicated chemisorption as involved in the mechanism. This suggests that the mechanism of fluoride up-take by AOCP is complex and involves both physical and chemical processes.

Table 3. 7 presents a comparison of the fluoride adsorption capacity of AOCP with capacities reported for other fluoride adsorbents, from which it is observed that at neutral pH of 7, AOCP has either a comparable or fairly higher adsorption capacity than some of the reported adsorbents.

Chapter 3 Drinking water defluoridation using aluminium (hydr) oxide coated pumice: Synthesis, equilibrium, kinetics and mechanism

101

Table 3.7. Comparison of AOCP capacity (q_{max}), with reported fluoride adsorbent capacities.

No	Adsorbent	q_{max}(mg/g), based on Langmuir isotherm model	pH	Reference
1	Bauxite	5.18	6	(Sujana & Anand, 2011)
2	High alumina content bauxite	3.15	7	(Lavecchia et al., 2012)
3	Fe(III)--Zeolite	2.31	7	(Sun et al., 2011)
4	Activated alumina	2.41	7	(Ghorai & Pant, 2005)
5	Activated alumina	1.45	7	(Ghorai & Pant, 2004)
6	Activated alumina	1.077	7	(Maliyekkal, 2006)
7	Manganese oxide coated alumina	2.85	7	(Maliyekkal, 2006)
8	Activated alumina	4.4	6.5 - 7	(Maliyekkal, 2008)
9	Manganese-amended activated alumina	10.12	6.5 - 7	(Maliyekkal, 2008)
10	**Aluminum oxide coated pumice(AOCP)**	**7.87**	**7**	**This work.**
11	Laterite	0.8461	7.5	(Sarkar, 2006)

3.3.7 Effect of solution pH on fluoride removal by AOCP

In most adsorption processes the initial solution pH is one of the important parameters that can significantly affect the extent of adsorption as well as determine the adsorption mechanism. This may be due to the effect of pH not only on the surface charge of the adsorbent, but also on the speciation and degree of ionization of the adsorbate (Wang et al., 2009; Li et al., 2011; Toor and Jin, 2012).

In the present work, adsorption of fluoride onto AOCP was studied at different pH ranging from 3 - 11.

The chemical equilibrium software package PHREEQC, was used for determining fluoride speciation at different solution pH (Appelo & Postma, 2005).

The adsorption of fluoride increased with increasing initial pH ($pH_{initial}$) in the acidic pH range, reached a maximum removal efficiency of 93.4% at $pH_{initial} = 6$ with a corresponding capacity of (0.43 mg/g) and thereafter declined with further increase in $pH_{initial}$ (Fig.3.9).

The variation of final pH (pH$_{final}$) at the end of batch adsorption experiment from the initial pH at the start of experiment is also shown in Fig.3.9.

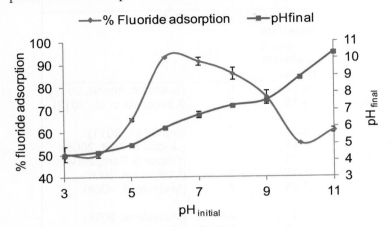

Fig. 3.9 Effect of pH on fluoride adsorption by AOCP

It was found from the XRF and EDX analyses that oxides of Al were a major component of AOCP. When the aluminum oxide particles are dry, the atoms at the surface differ from the atoms in the internal structure by being coordinately unsaturated. When the surfaces become hydrated as in aqueous environments, the coordinative unsaturation leads to a dissociative chemisorption of water which results in the formation of surface hydroxyl groups (Rosenqvist, 2002). Using an idealized structure of γ- Al$_2$O$_3$ to illustrate: when it is dry the top (surface) layer contains only oxide ions regularly arranged over aluminum ions in octahedral sites in the next lower layer as shown in Fig.3.10. The surface layer contains only half of the oxide ions present in the next lower layer, with aluminum ions located in all interstices between the oxide ions.

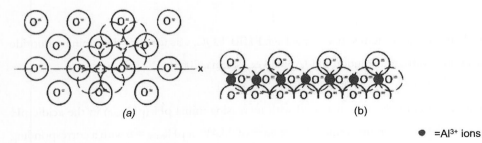

● =Al^{3+} ions

Fig. 3.10: Dry surface of γ- Al$_2$O$_3$: (a) top layer of only oxide ions, (b) perpendicular section (Source: Goldberg et al., 1996).

Chapter 3 Drinking water defluoridation using aluminium (hydr) oxide coated pumice: Synthesis, equilibrium, kinetics and mechanism

103

When the surface of the particles come into contact with water, the water is chemisorbed to convert the surface layer of oxide ions to a filled, square lattice of hydroxyl ions as shown in Fig. 3.11.

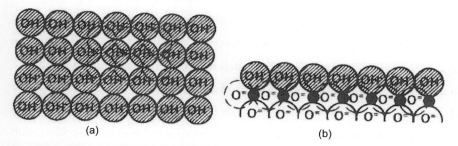

(a)

(b)

Fig. 3.11: Hydrated surface of γ- Al_2O_3: (a) top layer of aluminol groups, (b) section perpendicular to hydrated surface (Goldberg et al., 1996).

The surface hydroxyl ions, coordinated in various ways with aluminum cations, constitute the functional groups (i.e aluminol groups: AlOH) or active surface sites of aluminum oxides (Goldberg et al., 1996) and are responsible for the characteristic reaction of the AOCP surfaces with contaminants in aqueous solutions.

A pH-dependent surface charge develops at the AOCP aluminol surfaces sites by proton transfer as described by the following equations:

$$AlOH + H^+ \leftrightarrow AlOH_2^+ \tag{3.25}$$

$$AlOH \leftrightarrow AlO^- + H^+ \tag{3.26}$$

where AlOH, $AlOH_2^+$, AlO^- are the neutral, protonated and deprotonated sites on AOCP, respectively. At pH below the pH_{pzc} of AOCP (=5.2) the surface is expected to be mainly positively charged while negative charge is retained on the particles at pH above pH_{pzc}.

The mechanisms involved in the uptake of an anionic adsorbate such as fluoride ions by an adsorbent such as AOCP may be through specific and/or non-specific adsorption processes. The specific adsorption involves ligand/anion exchange reactions where the anions displace OH⁻ and/or H_2O from the adsorbent surface. The non-specific adsorption involves columbic forces of attraction between the anion and the adsorbent and depends mainly on the pH_{pzc} of the adsorbent (Li et al., 2011; Sujana et al., 2009).

The fluoride adsorption capacity of AOCP is expected to decrease at $pH \geq pH_{pzc}$ due to either the mainly neutral or negative charge of AOCP surface, if the removal mechanism was by non-specific adsorption only. It was, however, observed from Fig 3.9 that the highest fluoride adsorption occured at pH 6 which is higher than the pH_{pzc} of AOCP (pH_{pzc} = 5.2), indicating that high anion uptake occurs even though the AOCP surface may be electrically negative. This observation suggests that the mechanism of fluoride adsorption by AOCP may probably be due to a combination of specific and non-specific adsorption processes. The removal mechanism at $pH \leq pH_{pzc}$, which is in the acidic range, is presumably due to columbic attraction of fluoride by the surface positive charges (i.e, equation 3.27) and/or ligand exchange reaction of fluoride with surface hydroxyl groups of the solid adsorbent (i.e equation 3.28):

$$AlOH_2^+ + F^- \leftrightarrow AlOH_2^+\text{----}F^- \qquad\qquad (3.27)$$

$$AlOH_2^+ \, F^- \leftrightarrow AlF \; + H_2O \qquad\qquad (3.28)$$

There was an observed increase in the final pH_{final} in the acidic range after the adsorption experiment, which is consistent with the columbic or ligand exchange type adsorption mechanism as suggested. At $pH \geq pH_{pzc}$, the predominant mechanism for fluoride adsorption may be through ligand exchange type interactions between the fluoride ions and hydroxyl groups represented as equation (3.29):

$$AlOH + F^- \leftrightarrow AlF + OH^- \qquad\qquad (3.29).$$

It was further observed from Fig 3.9 that the lowest amounts of fluoride were removed under both acidic and alkaline conditions. The decrease in fluoride removal in the acidic range may be due to the formation of hydrofluoric acid (HF) according to the following reaction:

Chapter 3 Drinking water defluoridation using aluminium (hydr) oxide
coated pumice: Synthesis, equilibrium, kinetics and mechanism

105

$$HF \leftrightarrow H^+ + F^-$$
(3.30).

According to the fluoride speciation calculations using PHREEQC, HF was the predominant species under acidic pH conditions. Because HF is weakly ionized in solution at low pH values, the corresponding uptake of fluoride is reduced since a fraction of the fluoride ions become unavailable for adsorption. The formation of HF reduces the columbic forces of attraction between fluoride and AOCP surface sites, thus contributing to a lower fluoride uptake (Chen et al., 2011; Eskandarpour et al., 2008). The observed decrease of fluoride uptake under alkaline conditions is probably due to electrical repulsion between negatively charged AOCP surface sites (see equation 3.26) and fluoride ions and/or competition for active surface sites between fluoride ions and hydroxyl ions, both of which hinder fluoride uptake (Sujana et al., 2009). The competition for adsorbent sites by hydroxyl ions under alkaline conditions, as suggested, agrees well with the decrease of final solution pH (Fig 3.9).

AOCP was, however, found to exhibit good fluoride adsorption within the pH range 6-9 (Fig 3.9). This makes it potentially suitable for use in groundwater treatment, particularly in remote areas of developing countries as the need for pH adjustment with the associated capital cost and operational and maintenance challenges could possibly be avoided.

3.3.8 Effect of storage time on the fluoride removal performance of AOCP

It is unlikely that a freshly produced fluoride adsorbent may be used immediately for water defluoridation. In practice, the adsorbent may be stored for sometime prior to application. The effect of long-term storage on the performance performance AOCP for water treatment was assessed in batch adsorption experiments, using freshly produced AOCP, and AOCP stored for 7 and 12 months under normal room conditions (temp. = 20 °C).

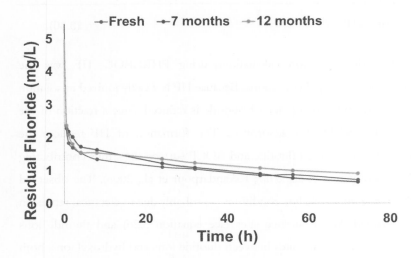

Fig. 3.12 Fluoride removal by AOCP stored at different times: adsorbent dose = 10 g/L, particle size = 0.8 - 1.12mm; model water: fluoride = 5 ± 0.2 mg/L, HCO_3 = 260 mg/L, pH = 7.0 ± 0.2.

The performance of AOCP was found to remain almost the same, even after 12 months of storage (Fig. 3.12), suggesting the aluminium (hydr)oxides were sufficiently bound onto the pumice paricle surfaces and also underwent no significant transformations under the storage conditions (room conditions), that could affect its efficiency.

3.4   Conclusions

Modification of pumice particle surfaces by aluminum oxide coating was found effective in creating hard surface sites for fluoride adsorption   in accordance with the HSAB concept. Aluminum oxide coated pumice (AOCP) reduced fluoride in model water from 5.0 ±0.2 mg/L to ≤ 1.5 mg/L (WHO guideline value). Thermal treatment of AOCP negatively affected its fluoride removal efficiency contrary to expectations, indicating thermal treatment of AOCP in its synthesis process may not augment for fluoride removal. Fluoride adsorption onto AOCP is complex, and both film and intra-particle diffusion contribute to the rate-determining step. The adsorption of fluoride by AOCP conformed reasonably to five isotherm models in the order: Generalized model > Langmuir type 2 > BET >Temkin >Dubinin-Radushkevich; with a maximum adsorption capacity of 7.87 mg/g.   The mechanism of fluoride removal by AOCP may probably be due to a combination of specific

Chapter 3 Drinking water defluoridation using aluminium (hydr) oxide
coated pumice: Synthesis, equilibrium, kinetics and mechanism

107

and non-specific adsorption processes. AOCP exhibited good fluoride adsorption within the pH range 6-9 which makes it potentially suitable for use in groundwater treatment, as the need for pH adjustment with the associated cost and operational difficulties especially in remote areas of developing countries could possibly be avoided. Long-term storage of AOCP had no significant effect on its fluoride removal performance. Based on results from batch kinetic experiments, it was observed that at a neutral pH of 7.0 ±0.1 which is a more suitable condition for groundwater treatment, the kinetics of fluoride adsorption by AOCP was quite comparable or perhaps fairly faster in the initial period of contact than that of activated alumina (AA) which is the commonly used adsorbent for water defluoridation. AOCP is thus promising and could also possibly be a useful fluoride adsorbent in developing countries where indigenous materials (i.e. pumice) could be used for possible cost reduction.

References

Abusafa, A and Yucel, H. 2002. Removal of ^{137}Cs from aqueous solutions using different cationic forms of a natural zeolite: clinoptilolite, Sep. Purif. Technol. 28 (2002) 103 - 116.

Alfara, A., Frackowiak, E., Beguin, F. 2004. The HSAB concept as a means to interpret the adsorption of metal ions onto activated carbon, Applied Sur. Sci. 228, 84-92.

Allen, S.J., Gan, Q., Matthews, R., Johnson, P.A. 2003. Comparison of optimised isotherm models for basic dye adsorption by kudzu. Bioresource Tech. 88, 143- 152.

Appelo, C.A and Postma, D. 2005. Geochemistry, groundwater and pollution, second ed. CRC Press, New York.

Ayoob, S., Gupta, A.K., 2008. Insights into isotherm making in the sorptive removal of fluoride from drinking water, J. Hazard. Mat., 152, 976 -985.

Bhargava D.S and D.J. Killedar, D.J. 1992. Fluoride adsorption on fishbone charcoal through a moving media adsorber, Water Res. 26(6), 781-788.

Biswas, K., Gupta, K., Ghosh, U.C. 2009. Adsorption of fluoride by hydrous iron (III)-tin(IV) bimetal mixed oxide from the aqueous solutions, Chem. Eng. J. 149, 196-206.

Bourika, K., Vakros, J., Kordulis, C., Lycourghiotis, A. 2003. Potentiometric mass titrations: Experimental and theoretical establishment of a new technique for determining of point of zero charge (PZC) of metal (hydr) oxides, J.Phys.Chem. B. 107, 9441-9451.

Chen, N., Zhang, Z., Feng, C., Zhu, D., Yang, Y., Sugiura, N. 2011. Preparation and characterization of porous granular ceramic dispersed aluminum and iron oxides as adsorbent for fluoride removal from aqueous solution, J. of Hazard. Mater. 186,863-868.

Das, N., Pattanaik, P., Das, R. 2005. Defluoridation of drinking water using activated titanium rich bauxite, J. of Colloid and Interface Sci., 292, 1- 10.

Dawood, S and Sen, T. K. 2012. Removal of anionic dye Congo red from aqueous solution by raw pine and acid-treated pine cone powder as adsorbent: Equilibrium, thermodynamics, kinetics, mechanism and process design, Water Res. 46, 1933-1946.

Eskandarpour, A., Onyango, M.S., Ochieng, A., Asai, S. 2008. Removal of fluoride ions from aqueous solution at low pH using schwertmannite. J. of Hazard. Mater. 152, 571-579.

Fan, X., Parker, D.J., Smith, M.D. 2003. Adsorption kinetics of low cost materials, Water Res.37, 4929-4937.

Chapter 3 Drinking water defluoridation using aluminium (hydr) oxide coated pumice: Synthesis, equilibrium, kinetics and mechanism

109

Faust, S.D and Aly, O.M. 1998. Chemistry of water treatment, second ed, CRC Press LLC, Florida.

Fawell, J., Bailey, K., Chilton, J., Dahi, E., Fewtrell, L., Magara, Y. 2006. Fluoride in drinking water, IWA Publishing, London.

Febrianto, J., Kosasih, A.N., Sunarso, J., Ju, Y., Indraswati, N., Ismadjia, S. 2009. Equilibrium and kinetic studies in adsorption of heavy metals using biosorbent: A summary of recent studies, J. of Hazard. Mater. 162, 616 -645.

Foo, F.Y and Hameed, B.H. 2010. Review: Insight into modeling of adsorption systems. Chem.Eng. J. 156, 2-10.

Ghorai, S., and Pant, K.K. 2005. Equilibrium, kinetics and breakthrough studies for adsorption of fluoride on activated alumina, Sep.and Purif. Technol. 2, 165-173.

Ghorai, S and Pant, K.K. 2004. Investigation on the column performance of lfuoride adsorption by activated alumina in a fixed-bed, Chem. Eng. J. 98, 165-173.

Goldberg, S., Davis, J.A., Hem, J.D. 1996. The surface chemistry of aluminium oxide and hydroxides, In: Sposito, G (Eds), The Environmental Chemistry of Aluminum. CRC Press, Florida.

Groen, J.C., Peffer, L.A.A., Perz-Ramirez, J. 2003. Pore size determination in micro-and mesoporous materials. Pitfalls and limitations in gas adsorption data analysis, Microporous and Mesoporous Mater. 60, 1-17.

Ho, Y.S., 2004. Selection of optimum sorption isotherm. Carbon 42, 2115e2116.

Jimenez-Nunez, M.L., Solache-Rios, M., Chavez-Garduno, J., Olguin-Gutierrez, M.T. 2012. Effect of grain size and interfering anion species on the removal of fluoride by hydrotalcite-compounds, Chem. Eng. J. 181-182, 371-375.

Kaneko, K. 1994. Determination of pore size and pore size distribution: Adsorbent and catalyst, J. of Memb. Sci. 96, 59-89.

Karadag, D., Koc, Y., Turan, M., Ozturk, M. 2007. A comparative study of linear and non-linear regression analysis for ammonium exchange by clinoptilolite zeolite, J. of Hazard. Mater. 144, 432 -437.

Kawasaki, N., Ogata, F., Takahasshi, K., Kabayama, M., Kakehi, K., Tanada, S. 2008. Relationship between anion adsorption and physicochemical properties of aluminum oxide, J. of Health Sci. 54 (3), 324-329.

Kinniburg, D. G. 1986. General adsorption isotherms. Environ. Sci. Technol. 20, 895-904.

Kiran, B and Kaushik, A. 2008. Cynobacterial biosorption of Cr(VI): Application of two parameter and Bohart Adams models for batch and column studies, Chem. Eng. J.144, 391-399.

Kumar, K.V. 2006a. Comparitive analysis of linear and non-linear method of estimating the sorption isotherm parameters for malachite green onto activated carbon. J. of Hazard. Mater. B136, 197-202.

Kumar, K.V. 2006b. Linear and non-linear regression analysis for the sorption kinetics of methylene blue onto onto carbon. J. of Hazard. Mater. B 137, 1538- 1544.

Kumar, K.V and Sivanesan, S. 2006. Isotherm parameters for basic dyes onto activated carbon: Comparison of linear and non-linear method. J. of Hazard. Mater.B129, 147-150.

Chapter 3 Drinking water defluoridation using aluminium (hydr) oxide coated pumice: Synthesis, equilibrium, kinetics and mechanism

111

Lavecchia, R., Medici, F., Piga, L., Rinaldi, G., Zuorro, A. 2012. Fluoride removal from water by adsorption on a high alumina content bauxite, Chem. Eng. Transactions. 26, part 1, 225-230.

Li, Y., Zhangm, P., Du, Q., Peng, X., Lui, T., Wang, Z., Xia, Y., Zhang, W., Wang, K., Zhu, H., Wu, D. 2011. Adsorption of fluoride from aqueous solution by graphene, J. of Colloid and Interface.Sci. 363, 348-354.

Liu, Y and Liu, Y. 2008. Biosorption isotherms, kinetics and thermodynamics. Sep. Pur. Tech. 61, 229-242.

M. Toor and B. Jin, Adsorption characteristics, isotherm, kinetics and diffusion of modified natural bentonite for removing daizo dye, Chem. Eng. J. 187 (2012) 79-88.

Mahvi, A.H., Heibati, B., Mesdaghinia, A., Yari, A.R. 2012. Fluoride adsorption by pumice from aqueous solution, E-Journal of Chem. 9(4), 1843 -1853.

Makov, G., 1995. Chemical Hardness in density functional theory. J. Phys. Chem., 99, 9337-9339.

Malakootian, M., Fatehizadeh, A., Yousef, N., Ahmadian, M., Moosazadeh, M. 2011. Fluoride removal using regenerated spent bleaching earth (RSBE) from groundwater: Case study on Kuhbonan water. Desalination, 277, 244-249.

Malakootian, M., Moosazadeh, M., Yousefi, N., Fatehizadeh, A. 2011. Fluoride removal from aqueous solutions by pumice: case study on Kuhbonan water, African J. of Environ. Sci. and Tech. 5(4) 299-306.

Maliyekkal, S.M., Sharma, A.K., Philip, L. 2006. Manganese-oxide-coated alumina: A promising sorbent for defluoridation, Water Res. 40, 3497-3506.

Maliyekkal, S.M., Shukla, S., Philip, L., Nambi, I.M. 2008. Enhanced fluoride removal from drinking water by magnesia-amended activated alumina granules, Chem. Eng. J. 140, 183- 192.

Mlilo, T., Brunson, L., Sabatini, D. 2010. Arsenic and fluoride removal using simple materials, J. Environ. Eng. 136(4),   391-398.

Moges, G., Zewge, F., Socher, M. 1996. Preliminary investigations on the defluoridation of water using fired clay chips, J. of African Earth Sci. 21(4), 479-482.

Mohapatra, D., Mishra, D., Mishra, S.P., Chaudhury, G.R., Das, R.P. 2004. Use of oxide minerals to abate fluoride from water, J. of Colloid and Interface Sci. 275, 355 - 359.

Nemr, A.E. 2009. Potential of pomegranate husk carbon for Cr(VI) removal from wastewater: kinetic and isotherm studies. J. of Hazard. Mater. 161, 132-141.

Nigussie, W., Zewge, F., Chandravanshi, B.S. 2007. Removal of excess fluoride from water using waste residue from alum manufacturing process, J. of Hazard. Mater.147, 954-963.

Onyango, M.S., Kojima, Y., Aoyi, O., Bernardo, E.C., Matsuda, H. 2004. Adsorption equilibrium modeling and solution   chemistry dependence of fluoride removal from water by trivalent-cation-exchanged Zeolite F-9. J. of Colloid and Interface Sci. 297, 341-350.

Ozacar, M and Sengil, I.A. 2005. A kinetic study of metal complex dye sorption onto pine sawdust, Process Bio.Chem. 40, 565-572.

Ozmihci, S and Kargi, F. 2006. Utilization of powdered waste sludge (PWS) for removal of textile dyestuffs from wastewater by adsorption. J. of Environ. Manage. 81, 307-314.

Chapter 3 Drinking water defluoridation using aluminium (hydr) oxide
coated pumice: Synthesis, equilibrium, kinetics and mechanism

113

Pearson, R.G., 1988. Absolute electronegativity and hardness: Application to inorganic chemistry. Inorg. Chem., 27, 734-740.

Pearson, R.G., 1993. The principle of maximum hardness. Acc.Chem.Res. 26, 25-255.

Pearson, R.G., 2005. Chemical hardness and density functional theory. J.Chem. Sci., 117(5), 369-377.

Reardon E.J., and Wang, Y. 2000. A limestone reactor for fluoride removal from wastewaters, Environ. Sci. Technol.34, 3247-3253.

Rosenqvist, J. 2002. Surface chemistry of Al and Si (hydr) oxide, with emphasis on nano-sized gibbsite (Al(OH)$_3$). Department of Chemistry, Inorganic Chemistry, Umea University, Sweden.

Sarkar, M., Banerjee, A., Pramanick, P.P., Sarkar, A.R. 2006. Use of laterite for the removal of fluoride from contaminated drinking water, J. of Colloid and Interface Sci. 302, 432-441.

Singh, S.K., Townsend, T.G., Mazyck, D., Boyer, T.H. 2012. Equilibrium and intra-particle diffusion of stabilized landfill leachate onto micro-and meso-porous activated carbon, Water Res.46, 491-499.

Sivasankar, V., Rajkumar, S., Murugesh, S., Darchen, A. 2012. Tamarind (*Tamarindus indica*) fruit shell carbon: A calcium-rich promising adsorbent for fluoride removal from groundwater, J. of Hazard. Mater. 225-226, 164 -172.

Sivasankar, V., Ramachandramoorthy, T., Darchen, A. 2011. Manganese dioxide improves the efficiency of earthenware in fluoride removal from drinking water, Desalination. 272, 179 - 186.

Sontheimer, H., Crittenden, J.C., Summer, R.S. 1998. Activated carbon for water treatment. DVGW-Forschungsstele, Engler–Bunte-Institut, Karlsruhe.

Sujana, M.G and Anand, S. 2011. Fluoride removal studies from contaminated groundwater by using bauxite, Desalination. 267, 222-227.

Sujana, M.G., Soma, G., Vasumathi, N., Anand, S. 2009. Studies on fluoride adsorption capacities of amorphous Fe/Al mixed hydroxides from aqueous solutions. J. of Fluorine Chem.130, 749-754.

Sun, Y., Fang, Q., Dong, J., Cheng, X., Xu, J. 2011. Removal of fluoride from drinking water by natural stilbite zeolite modified with Fe (III), Desalination. 277, 121-127.

Toor, M and Jin B. 2012. Adsorption characteristics, isotherm, kinetics and diffusion of modified natural bentonite for removing daizo dye. Chem. Eng. J. 187, 79-88.

Tseng, R.L. 2006. Mesopore control of high surface area NaOH-activated carbon, J. of Colloid and Interface Sci. 303, 494- 502.

Tseng, R.L., Tseng, S.K. 2006. Characterization and use of high surface area activated carbons prepared from cane pith for liquid-phase adsorption, J. of Hazard. Mater. B136, 671- 680.

Vigneresse, J. L., 2012. Chemical reactivity parameters (HSAB) applied to magma evolution and ore formation. Lithos, 153, 154-164.

Vigneresse, J.L. 2008. Evaluation of the chemical reactivity of the fluid phase through hard-soft acid-base concepts in magmatic intrusions with applications to ore generation, Chem. geol. 263, 69-81.

Vigneresse, J.L., Duley, S., Chattaraj, P.K., 2011. Describing the chemical characteristic of a magma. Chem. Geol., 287, 192-113.

Wang, S., Ma, Y., Shi, Y., Gong, W. 2009. Defluoridation performance and mechanism of nano scale aluminum oxide hydroxide in aqueous solution, J. Chem.Technol. Biotechnol. 84, 1043-1050.

Chapter 3 Drinking water defluoridation using aluminium (hydr) oxide
coated pumice: Synthesis, equilibrium, kinetics and mechanism

115

Weber, W.J and Morris, J.C. 1963. Intraparticle diffusion during the sorption of surfactants onto activated carbon, J. of Sanitary Eng. Div, AESC.89, 53-61.

WHO, Guidelines for drinking water quality incorporating 1st and 2nd addenda, recommendations, Third ed. World Health Organization, Geneva, Switzerland, 2008.

Yousef, R.I., El-Eswed, B., Al-Muhtaseb, A. 2011. Adsorption characteristics of natural xeolites as solid adsorbents for phenol removal from aqueous solution: Kinetics, mechanism, and thermodynamicss studies, Chem. Eng. J. 171, 1143-1149.

Zafar, M.N., Nadeem, R., Hanif, M.A. 2007. Biosorption of nickel from protonated rice bran, J. of Hazard. Mater. 143, 478 -485.

4

Laboratory-scale column filter studies for fluoride removal with aluminum (hydr) oxide coated pumice, regeneration and disposal

Main part of this chapter has been prepared as draft journal article for publication in Environmental Science and Technology.

Chapter 4 Laboratory-scale column filter studies for fluoride removal with
aluminum (hydr) oxide coated pumice, regeneration and disposal

117

Abstract

The efficacy of aluminum (hydr) oxide coated pumice (AOCP) for water defluoridation, has been assessed and demonstrated under batch experimental conditions. The aims of the study reported in this chapter were to; (i) investigate the suitability of AOCP under dynamic conditions in laboratory-scale column experiments, which is a more appropriate approach for generating useful information for the design of full scale water treatment systems and, (ii) explore the potential for the regeneration of exhausted AOCP for re-use, as that can contribute to its economic viability. AOCP was capable of reducing model water with fluoride concentration of 5.0 ± 0.2 mg/L to ≤ 1.5 mg/L (WHO guideline), under continuous flow conditions. An empty bed contact time (EBCT) of 24 min was found a suitable guide contact time for design purposes, with respect to optimal use of AOCP fluoride adsorption capacity. Using Pearson's chi-square goodness-of-fit method, the Thomas and Adams-Bohart models were found capable of describing the breakthrough experimental data, and also the bed depth service time (BDST) model, based on the coefficient of determination (R^2). In contrast to the regeneration of most adsorbents, it was found that the fluoride adsorption capacity of exhausted AOCP after the first cycle of regeneration was not only fully (100%) restored, but increased by more than 30% (compared to that of fresh AOCP) under batch conditions and more than 50% under continuous flow conditions, suggesting the usefulness of the regeneration concept/approach. This may presumably contribute to the economic viability of AOCP for water defluoridation.

4.1 Background

Batch and continues flow adsorption studies are two common techniques employed to assess the efficacy of a prospective adsorbent. The efficacy of aluminum (hydr)oxide coated pumice (AOCP) for water defluoridation was assessed through batch adsorption experiments, and was found to be promising and could possibly be a useful fluoride adsorbent in developing countries, especially where pumice is locally available (as presented in chapter 3). Batch adsorption experiments are relatively quick and easy methods applied for establishing adsorption isotherms, and are useful as preliminary experiments for evaluation and comparison of performances of different adsorbents. Even though isotherms from batch experiments provide important information for designing of sorption systems, the results produced from such experiments are not as informative as that from column experiments. Isotherms are outcomes of static equilibrium test and several limitations affect the extrapolation of isotherm data to operational parameters of a full scale continues flow system based on a given adsorbent (Shih, et al, 2003; Quintelas et al., 2013). Laboratory-scale column studies on the other hand, simulate the dynamics of a fixed bed reactor that can be used to generate useful parameters, based on analysis of experimental breakthrough curves, which can be employed for design of full-scale water treatment systems.

Additionally, one of the factors that may contribute to the economic viability of any adsorbent is its potential for regeneration for reuse (Maliyekkal et al., 2006; Sun et al., 2011; Maliyekkal et al., 2008; Chauhan et al., 2007; Tresintsi et al.,2014;). Moreover, spent fluoride adsorbent if not disposed of into the environment safely, could become an important source of fluoride for groundwater re-contamination. For AOCP, however, studies related to its regeneration when exhausted and the safe disposal of spent AOCP into the environment has not yet been conducted.

The aims of the present study were therefore to: (1) investigate the fluoride removal behavior of AOCP under dynamic conditions using a laboratory scale fixed-bed adsorption column, (2) investigate the effects of bed depths and empty bed contact times (EBCTs) on the performance of AOCP and determine the contact time for an optimal use of the

Chapter 4 Laboratory-scale column filter studies for fluoride removal with
aluminum (hydr) oxide coated pumice, regeneration and disposal

119

adsorbent capacity, (3) explore a simple and innovative method for regeneration of exhausted AOCP, (4) explore a simple treatment method for stabilizing spent AOCP to reduce the risk of environmental pollution with fluoride, when disposed of.

The defluoridation performance of the regenerated exhausted AOCP (RAOCP) was studied under similar experimental conditions as that of the freshly synthesized AOCP. Adsorption isotherms and FTIR spectroscopic analysis were employed to gain an insight in the mechanism involved in the adsorption of fluoride onto RAOCP.

4.2 Material and methods

4.2.1 Preparation of adsorbent

AOCP used for the fixed-bed adsorption experiment was synthesized using similar procedure described in Chapter 3. However, to produce relatively larger quantities of AOCP required for the column experiments, a sufficient amount of 0.5 M $Al_2(SO_4)_3$ was added to completely soak about 3 kg of dried pumice in 20 L plastic bucket containers, well stirred and drained. This was dried and subsequently soaked in 3M NH_4OH to neutralize and complete the coating process as described in the previous chapter.

4.2.2 Fluoride model water

Fluoride model water used for the fixed-bed adsorption experiments was prepared under continuous flow conditions in a 150 L model water tank. Model water was prepared, using Delft (The Netherland) tap water with composition as indicated in chapter 3 in order to simulate natural water composition. The model water tank was continuously fed with the tap water (at a flow rate of 235 ml/min), with a continuous injection of stock solutions of fluoride (276 mg/L), bicarbonate (26,950 mg/L) and HCl (0.5 mg/L) at dosing rates of 5 ml/min, 5 ml/min and 1.75 ml/min, respectively, using small capacity dosing pumps (MASTERFLEX, Cole Parmer, USA). The tap water and the dosed stock solutions were completely mixed in the model water tank, with a mechanical stirrer, to obtain a final influent fluoride concentration (C_o(mg/L) of 5.0 ± 0.2 and bicarbonate concentration of

330.0 ± 5 mg/L, similar to that found in groundwater in the Northern region of Ghana. A neutral pH of 7.0 ± 0.2 was used for the experiment.

4.2.3 Column experiments

Laboratory-scale fixed bed column experiments were performed using a PVC pipe of 60 mm internal diameter and length of 3.2 m. The schematic diagram of the column experimental set-up is presented in Fig. 4.1. The column was packed with AOCP (and subsequently with RAOCP after regeneration of exhausted AOCP) to a total bed depth of 2.5 m. The influent fluoride solution from the model water tank was passed through the packed bed column in a down-flow mode at a design flow (Q_d) of 14.1 L/h (equivalent to a filtration rate of 5.0 m/h). The fixed bed column experiments were employed to evaluate the effects of different bed depths and empty bed contact times on the breakthrough curves. Thus water samples were taken at bed depths; 0.5 m, 1 m, 1.5 m, 2 m and 2.5 m, from the top of the filter bed (Fig. 4.1). These depths also corresponded to empty bed contact times (EBCT) of 6, 12, 18, 24 and 30 min, respectively. Samples of filtrate from the different bed depths were collected at pre-determined time intervals and analyzed for the residual fluoride concentrations (C_t(mg/L)).

Fourier-transform infrared (FTIR) spectra of RAOCP (a) before and (b) after fluoride adsorption, were collected using Nicolet 6700 FTIR spectrophotometer (Thermo Instrument, USA).

4.2.4 Regeneration of exhausted AOCP and fluoride adsorption studies

The HSAB concept was explored as a simple but innovative approach in the regeneration process, aimed at creating more active sites in order to increase the restored capacity of the regenerated adsorbent.

Regeneration was accomplished by re-coating the surfaces of the fluoride-saturated (exhausted) AOCP (FAOCP) with a new layer of aluminum (hydr)oxides, creating new active sites for restoring the fluoride adsorption capacity. Recoating was performed using similar procedure applied for the preparation of fresh AOCP (chapter 3). In contrast, however, the influence of the adsorbed fluoride on the FAOCP-Al coating solution interfacial properties (hence the coating reaction/mechanism) was explored as a simple approach in the regeneration step, aimed at incorporating more Al^{3+} in accordance to the HSAB concept, hence, creat higher number of active sites in order to increase the restored capacity of the regenerated adsorbent. Thus after running the column experiment with AOCP till full saturation, the fluoride-saturated AOCP (FAOCP) was removed and air-dried. The adsorbed fluoride on the FAOCP particle surfaces were not stripped of (generally in most regeneration processes, this is stripped of), but re-soaked in sufficient amount of 0.5M $Al_2(SO_4)_3$ solution and the re-coating procedure followed to accomplish the regeneration/restoration. The regenerated adsorbent (RAOCP) was subsequently used for fluoride adsorption studies in similar continuous flow column experiment as conducted with fresh AOCP, and also in batch equilibrium studies using similar conditions as described for fresh AOCP (chapter 3).

Aluminum content of RAOCP sample was extracted by acid digestion at a temperature of 200°C, using reagent grade concentrated HNO_3.

4.2.5 Stabilization of spent (waste) AOCP and leaching test for safe disposal

For the purpose of disposing off when it can no longer be used, spent (waste) AOCP was re-coated with a final layer of Al as a possible simple treatment method for its stabilization that could possibly allow a safe disposal. The option of disposing off the spent AOCP without any treatment and without a risk of polluting the environment with fluoride was also assessed, as that could be a cheaper disposal option.

The toxicity characterization leaching procedure (TCLP), developed by the United States Environmental Protection Agency (US-EPA, 1992), was employed to characterize the leaching properties of the spent AOCP with and without its stabilization, in order to assess the risk of potential environmental mobility of fluoride contained in these waste materials, and determine if any has the characteristics of hazardous materials or otherwise. For the

TCLP, two extraction liquids were used in the leaching test: liquid 1 was CH_3COOH with a pH of 4.9 $\pm$ 0.05 and liquid 2 was CH_3COOH with pH 2.9 $\pm$ 0.05. The extraction liquid used for any of the waste (spent) adsorbent materials (i.e with and without stabilization) was determined based on the material alkalinity. The pH values of the spent AOCP samples were determined in Milli-Q water using a 1: 2.5 ratio (spent AOCP: Milli-Q water). Liquid 1 was used for the stabilized spent AOCP samples with pH< 5 (pH = 4.7) and liquid 2 was employed for spent AOCP without stabilization with pH > 5 (pH = 7.4). The amount of extraction liquid was 20 times the weight of spent AOCP, and extraction was carried out for 18 h. The spent AOCP samples were transferred into polyethene bottles containing the extraction liquid and mounted on a shaker for 18 h. Samples of the liquid extract were filtered through 0.7 μm glass fiber filter in order to separate the liquid from the solid phase. The liquid extracts were then analyzed for the fluoride content and compared against thresholds established by the US-EPA of 100 times National Interim Primary Drinking Water Standard (US-EPA, 1992; Margui et al, 2004). All leaching test were done in triplicate to check reproducibility.

Chapter 4 Laboratory-scale column filter studies for fuoride removal with
aluminum (hydr) oxide coated pumice, regeneration and disposal

123

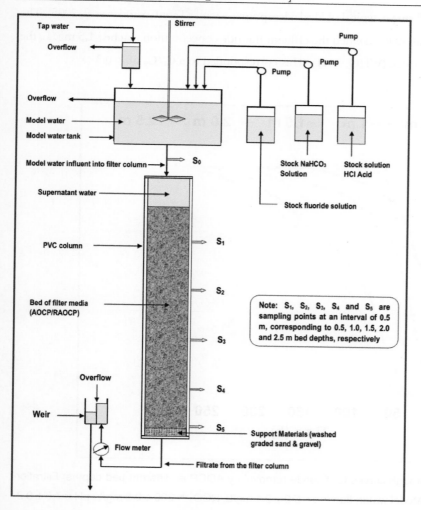

Fig 4.1 Schematic diagram of column experimental set-up

4.3 Results and Discussions

4.3.1 Breakthrough curves for fluoride adsorption onto AOCP

The results of fluoride adsorption onto AOCP under the continuous flow conditions are presented in the form of breakthrough curves for the five different bed depths as shown in Fig. 4.2. The breakthrough curves, expressed as the ratio of effluent to influent fluoride concentration (C_t/C_o) as a function of time, depicts the loading behavior of fluoride ions from solution onto the AOCP column filter. The curves were observed to follow the characteristic "S" shape, typical of ideal adsorption systems (Taty-Costodes et al, 2005). The breakthrough was attained when the effluent fluoride concentration reached 1.5 mg/L, the WHO guideline value (WHO, 2011). This corresponded to a C_t/C_o of ≈ 0.3

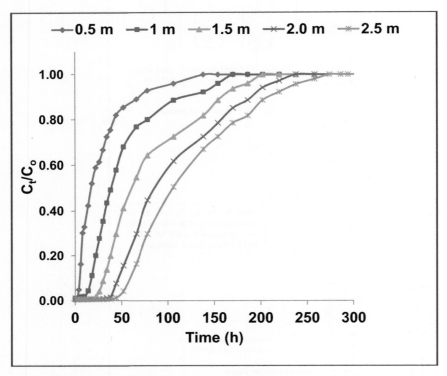

Fig 4.2 Breakthrough curves for fluoride removal by AOCP at different bed depths: Filtration rate = 5.0 m/h; Model water: fluoride = 5.0 ± 0.1 mg/L, HCO_3^- = 330 ± 0.5 mg/L, pH = 7.0 ± 0.2, temp. = 20 °C.

4.3.2 Effect of bed depth on breakthrough

The breakthrough times (t_b) or service times in hours when the effluent concentration reached 30% of the influent fluoride concentration and the treated volumes of water for different bed depths are given in Table 4.1. The volumes of water treated until breakthrough were observed to increase with increased bed depth from 0.5 to 2.5 m. The variation of volume of treated water with filter bed depth may be attributed to factors, including: the mass transfer phenomenon that takes place in the adsorption process at lower (i.e

Chapter 4 Laboratory-scale column filter studies for fluoride removal with aluminum (hydr) oxide coated pumice, regeneration and disposal

125

shallower) bed depths, and the variation of contact times with bed depth (Taty-Codes et al., 2005; Chen et al., 2011). At shallower bed depths (Fig. 4.1), axial dispersion phenomenon presumably dominates the mass transfer process and reduces the diffusion of fluoride ions into the mass of AOCP. Moreover at shallower bed depths which correspond to shorter empty bed contact times (Table 4.1), the fluoride ions do not have enough time to diffuse deeper into the whole of the AOCP mass for increased uptake. These factors most likely contributed to a reduction in the breakthrough volume of treated water from 1.0998 m^3 to 0.1128 m^3 as observed, when the AOCP bed depth decreased from 2.5 m to 0.5 m.

The fluoride adsorption capacities of AOCP at breakthrough (q_b) (mg/g) for given bed depths were estimated using equation (4.1), where M is the adsorbent mass for a given bed depth:

$$q_b = (Q_d t_b C_o)/M$$

(4.1).

The fluoride adsorption capacities (Table 4.1) were also observed to increase with increasing bed depth. This was due to an increase in AOCP mass at deeper bed depths, which provided more surface area and binding sites for more fluoride uptake and hence resulted in increased bed volumes (BV) of water treated (Table 4.1). The adsorption capacities and number of bed volumes of water treated, however, increased with the bed depth till 2 m, where a maximum adsorbent capacity of 2.33 mg/g and a corresponding maximum of 165 bed volumes were attained, and thereafter declined with further increase of bed depth (Table 4.1). At a constant fluoride concentration in feed water and constant flow rate, the bed depth at which the adsorption capacity began to decline presumably represent a point where there was an increasing AOCP dose for a fixed solute load, as that may result in the availability of fewer fluoride ions per unit mass of AOCP, i.e., lesser fluoride/AOCP ratio with increasing adsorbent dose (Taty-Codes et al., 2005; Zheng et al., 2008; Sulaiman et al, 2009; Chen et al., 2011; Salifu et al, 2013; Garcia-Sanchez et al., 2013).

Table 4.1: Bed depth, breakthrough times, volume of water treated and adsorption capacities for fluoride removal in AOCP column filter

Bed depth (m)	EBCT (min.)	Mass of AOCP (M_{AOCP}) (g)	Volume of AOCP (m³)	Breakthrough time (t_b) (h)	Volume of water treated (m³)	No. of bed volumes (BV)	AOCP usage rate (g/L)	Fluoride adsorption capacity q_b (mg/g)
0.5	6	499	0.0014	8	0.1128	80	4.42	1.128
1.0	12	998	0.0028	26	0.3666	130	2.72	1.834
1.5	18	1497	0.0042	44	0.6204	146	2.41	2.068
2.0	24	1996	0.0057	66	0.9306	165	2.14	2.326
2.5	30	2495	0.0071	78	1.0998	156	2.27	2.199

4.3.3 Effect of EBCT on AOCP performance and usage rate

Two of the important design parameters that need to be determined for a fixed bed adsorption treatment system are the EBCT and adsorbent usage rate (i.e., mass of adsorbent/volume of water treated before breakthrough). The EBCT affects the bed life (number of bed volumes (BV) of water treated before breakthrough) of the system, and its proper selection is required to fully utilize the adsorbent capacity (Faust and Ally, 1998; Thomas and Crittenden, 1998; Shih et al., 2003). The effect of EBCT on AOCP usage rate, which determines how often the adsorbent must be regenerated or replaced and the corresponding bed life, were examined using EBCTs of 6, 12, 18, 24 and 30 min, and the results are presented in Table 4.1 and Fig. 4.3.

From Table 4.1 and Fig. 4.3, an initial rapid decrease of AOCP usage rate was observed with increasing EBCT from 6 to 24 min, where a minimum value of 2.14 g/L was attained, and thereafter started to increase. Correspondently, the bed life also increased with increase of EBCT till the maximum value of 165 BV was attained at an EBCT of 24 min, and thereafter began to decrease. Thus 24 min was considered an EBCT for optimizing the use of AOCP adsorption capacity and which could be useful guide for the design of full scale applications, base on the design flow (Q_d) applied in this study .

Chapter 4 Laboratory-scale column filter studies for fluoride removal with aluminum (hydr) oxide coated pumice, regeneration and disposal

127

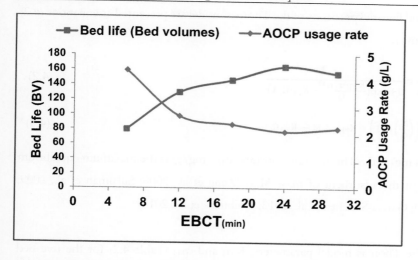

Fig. 4.3 Effect of EBCT on AOCP performance and usage rate

4.4 Modeling of breakthrough profiles

The use of breakthrough curve models for predictions of the effects of various operational parameters (e.g. flow rate, influent adsorbate concentration) on the breakthrough profile for an adsorbate effluent concentration in order to determine the optimal operational conditions, is a useful approach for good design of fixed-bed adsorption treatment systems (Taty-Costodes et al., 2005; Zheng et al., 2008; Han et al., 2009; Sulaiman et al., 2009; Chen et al., 2011). The column experimental data were therefore fitted to three commonly used mathematical models, namely, the Thomas, the Adams-Bohart and the bed depth service time (BDST) models for describing the dynamic behavior of fluoride adsorption in an AOCP fixed-bed column filter.

4.4.1 Thomas model

Derivation of the Thomas model assumes: (i) Langmuir kinetics of adsorption-desorption with no axial dispersion, and (ii) that the rate driving force obeys the second-order reversible reaction kinetics.

The non-linear and linear expressions of the Thomas model are presented as equations (4.2) and (4.3), respectively:

$$\frac{C_t}{C_o} = \frac{1}{1+\exp\left(\frac{K_{TH}q_{TH}M}{Q}-K_{TH}C_ot\right)} \qquad (4.2)$$

$$\ln\left(\left(\frac{C_o}{C_t}\right)-1\right) = K_{TH}q_{TH}M - K_{TH}C_ot \qquad (4.3)$$

where K_{TH} (L/h mg) is the Thomas rate constant, q_{TH} (mg/g) is the maximum equilibrium uptake capacity of the adsorbent of mass M(g) (Zeng et al., 2008; Sulaiman et al., 2009; Hans et al., 2009; Garcia-Sanchez et al., 2013; Quintelas et al., 2013).

The values of the Thomas model parameters, K_{TH} and q_{TH} (Table 4.2) for the five bed depths were determined using non-linear optimization techniques based on equation 4.2. The goodness of fit of the Thomas model to the experimental breakthrough data was evaluated using the Pearson's Chi-squared (χ^2) goodness-of-fit-method. The χ^2 values were calculated from equation 4.4.

$$X^2 = \sum_{i=1}^{i=N} \frac{\left(C_{t(exp.)}-C_{t(Cal.)}\right)^2}{C_{t(cal.)}} \qquad (4.4)$$

where $C_{t(cal)}$ and $C_{t(exp.)}$ represent the calculated and measured concentration, respectively, at time t (h) (Allen et al., 2003; Ho et al., 2004; Physick et al., 2016).

Fig.4.4 shows a comparison of the experimental and simulated breakthrough curves based on the Thomas model on the same axial setting (shown for three representative bed depths), which generally showed good correlations for the different bed depths, except some few discrepancies. The non-linear χ^2 values, indicated in Table 4.2, were observed to be generally small at all bed depths. The degrees of freedom (d_f) (= number of experimental data points, minus number of model parameters) were ≥ 27 at all bed depths, and the respective χ^2 values ($0.2869 - 0.783$) were found to be less than their corresponding critical values; ($\alpha = 0.05$, $d_f = 27$, critical value of $\chi^2 = 40.113$), where α is the significance level. This indicates there were statistically no significant differences between the experimental data

Chapter 4 Laboratory-scale column filter studies for fuoride removal with
aluminum (hydr) oxide coated pumice, regeneration and disposal

129

points and the calculated data by the Thomas model for all bed depths, suggesting the model was adequate in describing the experimental breakthrough curves (Allen et al., 2003; Ho et al., 2004; Physick et al., 2016). The values of the maximum equilibrium uptake capacity obtained from the Thomas model, q_{TH}, increased with increasing bed depth and therefore followed a similar trend as observed for the variation of volumes of treated water with increase of bed depth (section 4.3).

The observed discrepancies between the experimental data and the calculated data from the Thomas model, may be due to the primary weakness inherent in the model. Derivation of the Thomas model excludes external (film) resistance and intra-particle mass transfer resistance in the adsorption mechanism, thus suggesting adsorbate–adsorbent surface reactions control the rate of adsorption, hence the breakthrough. Most adsorption processes are, however, usually not limited by surface reaction kinetics, but often controlled by external and/or intra-particle mass transfer (Aksu and Gönen, 2004; Ghorai and Pant, 2005). The observed discrepancies in this study therefore suggests external and/or intra-particle mass transfer may be the rate controlling steps in the adsorption of fluoride onto AOCP in the column filters. This is also consistent with the kinetic study of the fluoride-AOCP system under batch conditions presented in the previous chapter of the thesis.

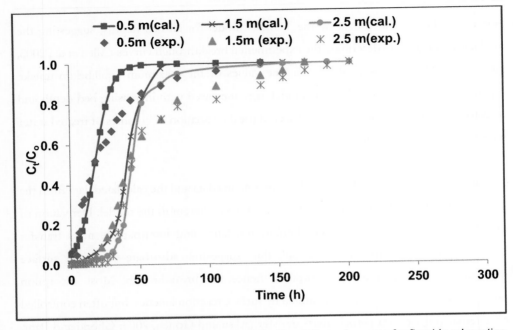

Fig.4.4 Comparison of experimental and simulated breakthrough curves for fluoride adsorption on AOCP based on the Thomas model.

Table 4.2 Thomas and Adams-Bohart model parameters at different bed depths for AOCP column.

Bed depth (m)	Thomas model			Adams-Bohart model		
	K_{TH}	q_{TH}	X^2	K_{AB}	N_o	X^2
0.5	0.0321	2.4914	0.7833	0.0527	705504.83	0.2069
1.0	0.0227	2.6154	0.2549	0.0277	870897.35	0.0814
1.5	0.0162	2.7573	0.2849	0.0163	1005877.31	0.0934
2.0	0.1184	3.0087	0.2993	0.0114	1120272.91	0.1191
2.5	0.0094	3.0992	0.2869	0.0112	1000182.48	0.0824

4.4.2 Adams-Bohart model

The Adams-Bohart model is employed for describing the initial part of a breakthrough curve with the validity of the model being limited to the concentration range of $\frac{C_t}{C_o} < 0.5$ (Garcia-Sanchez et al., 2013). The model assumes the rate of adsorption is limited by external (film) mass transfer. The non-linear form of the model is given as equation (4.5), which can be linearized as equation (4.6).

Chapter 4 Laboratory-scale column filter studies for fluoride removal with
aluminum (hydr) oxide coated pumice, regeneration and disposal

131

$$\frac{C_t}{C_o} = exp\left(K_{AB}C_ot - K_{AB}N_o\frac{Z}{v}\right) \tag{4.5}$$

$$ln\left(\frac{C_t}{C_o}\right) = K_{AB}C_ot - K_{AB}N_o\frac{Z}{v} \tag{4.6}$$

where K_{AB} (L/mg h) is the kinetic constant, v (m/h) is the linear flow rate, Z (m) is a given column bed depth, N_o (mg/L) is the saturation concentration, and time t (h) ranges from the start to the fluoride breakthrough point. The linear flow rate was calculated from equation (4.7).

$$v = \frac{Q}{A} \tag{4.7}$$

where Q (m^3/h) is the volumetric flow rate, and A (m^2) is the cross sectional area of the bed (Garcia-Sanchez et al., 2013; Han et al., 2009; Quiteles et al., 2013). The Adams-Bohart model parameters, K_{AB} and N_o, presented in Table 4.2, were also determined by non-linear regression analysis according to equation (4.5).

Similar application of Pearson's Chi-squared (χ^2) goodness-of-fit-method suggested the suitability of the Adams-Bohart model for describing the initial sorption process of fluoride under continues flow conditions in the AOCP column filters. A similar comparison of the plots of the experimental and the calculated breakthrough curves, based on the Adams-Bohart equation, generally showed very good correlations for all bed depths (Fig.4.5) (shown for three representative depths). Only marginal discrepancies were observed at the lower bed depths (0.5 and 1.0 m). The applicability of the model also supports the contribution of external mass transfer to the rate limiting step of the fluoride adsorption process (Aksu and Gönen, 2004; Quintelas et al., 2013). The validity of the model is, however, limited to the range of concentrations used (i.e. $\frac{C_t}{C_o} < 0.5$).

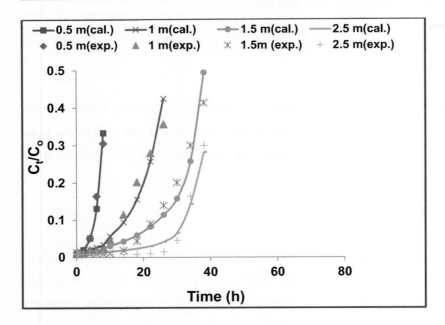

Fig. 4.5 Comparison of simulated and experimental breakthrough curves for different bed depths based on the Adams-Bohart model.

4.4.3 Bed depth service time (BDST) model

In developing the BDST model, it was supposed that: (i) intraparticle diffusion and external mass transfer are negligible, and, (ii) the adsorption kinetics is controlled by surface chemical reaction between adsorbate in solution and unused adsorbent capacity.

The BDST model can be expressed as equation (4.8):

$$t = \frac{N_{bdst}}{C_o v}Z - \frac{1}{K_{ads}C_o}\ln\left(\frac{C_o}{C_b} - 1\right)$$

(4.8),

where C_b (mg/L) is the breakthrough fluoride concentration (1.5 mg/L), N_{bdst} (mg/L) is the dynamic bed adsorption capacity, t (min) are the service times for different bed depths till breakthrough, K_{ads} (L/mg min) is the adsorption rate constant and v (cm/min), the linear flow rate (Taty-Costodes et al., 2005; Han et al., 2009; Chen et al., 2011; Garcia-Sanchez et al., 2013).

Chapter 4 Laboratory-scale column filter studies for fuoride removal with aluminum (hydr) oxide coated pumice, regeneration and disposal

133

The BDST model parameters, N_{bdst} and K_{ads}, were determined from the slope and intercept, respectively, of the linear plot of t versus Z (cm) for the AOCP experimental data, based on equation (4.8), and is presented in Fig.4.6. (The BDST plot for RAOCP experimental data is also included in Fig. 4.6, and is further discussed in section 4.5.2).

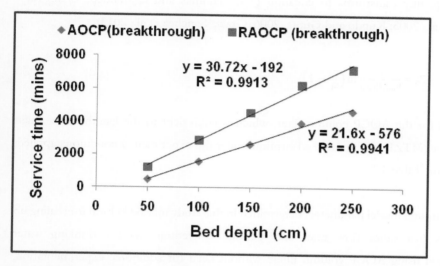

Fig 4.6 Service time vs. bed depth at breakthrough for AOCP and regenerated AOCP (RAOCP).

The model parameters and the coefficient of determination (R^2) are given in Table 4.3. The high R^2 value indicates the BDST model is also appropriate for describing the adsorption of fluoride onto AOCP under the continuous flow conditions.

The slope, m (min/cm) of the BDST line indicates the time required for the mass transfer zone (MTZ) to travel a unit length through the adsorbent bed and it is useful for predicting the bed performance (Taty-Costodes et al., 2005; Gorai and Pant, 2005). For the AOCP-fluoride system under the applied continuous flow conditions, m, was found to be 21.6 (min/cm) (Table 4.3).

Table 4.3: BDST model parameter for fluoride adsorption onto AOCP.

Slope (min/cm)	N_{bdst} (mg/L)	K_{ads} (L/mg min)	Z_o (cm)	R^2
21.6	895.4782	0.000294	26.67	0.9941

The theoretical or critical adsorbent bed depth (Z_o) required for preventing the fluoride concentration from exceeding the treatment target, C_b is obtained when t = 0, and equation (4.8) consequently transforms to equation (4.9) (Thomas and Crittenden, 1998; Taty-Costodes et al., 2005; Kundu and Gupta, 2006; Gracia-Sanchez et al., 2013).

$$Z_o = \frac{v}{K_{ads}N_{bdst}} \ln\left(\frac{C_o}{C_b} - 1\right) \qquad (4.9)$$

Z_o obtained for the AOCP column filter, which is equivalent to the length of the mass transfer zone (MTZL) for fluoride adsorption under the experimental conditions applied, was 26.67 cm (Table 4.3).

The breakthrough model parameters determined in this study may be helpful for scaling up the process (for other flow rates and fluoride concentrations) for drinking water defluoridation using AOCP column filters without need for additional experimentation, which could possibly save time and cost.

4.5 Fluoride adsorption performance of RAOCP and comparison with that of AOCP

The fluoride removal performance of the regenerated AOCP (RAOCP) was evaluated under similar batch and column experimental conditions, and the effectiveness of the regeneration procedure was assessed based on the degree of recovery of the adsorption capacity. Other performance measures used for comparing the effectiveness or efficiency of AOCP before and after regeneration included the adsorbent usage rate, number of bed volumes (BV) of water treated before breakthrough, the critical bed depth (Z_o) or MTZL and the time required for the MTZ to travel a unit length through the adsorbent depth (Taty-Costodes et al., 2005; Ghorai and Pant, 2006; Kundu and Gupta, 2006; Maliyekkai, 2006; Hendricks, 2011).

4.5.1 Batch equilibrium fluoride uptake by RAOCP

The equilibrium experimental data for fluoride removal with RAOCP were fitted to Langmuir type 4 and the Freundlich isotherm models (Maliyekkai, 2006; Chen et al., 2011), for estimating the adsorption capacity. The equation of the Langmuir type 4 is indicated in Table 3.5 of chapter 3, and the linear form of the Freundlich model is represented as equation 4.10.

$$\log q_e = \log K_F + \frac{1}{n} \log C_e \qquad\qquad 4.10$$

where K_F (mg/g)(L/mg)$^{1/n}$ and n are the Freundlich constants indicative of adsorption capacity and adsorption intensity, respectively, C_e (mg/L) is the equilibrium concentration of fluoride in aqueous solution and q_e (mg/g) the amount of fluoride removed onto RAOCP at equilibrium. A plot of log q_e against log C_e yields a straight line of slope 1/n and intercepts K_F (Chen et al., 2011; Maliyekkai, 2006).

The Langmuir and Freundlich isotherm plots of the equilibrium data for fluoride uptake by RAOCP are shown as Fig.4.7 (a) and (b), respectively. The isotherm constants, calculated from the slopes and intercepts of the linear plots of the equilibrium data according to Langmuir and Freundlich equations, respectively, are presented in Table 4.4. The high R^2 values indicate the two models were able to reasonably describe the equilibrium uptake of fluoride onto RAOCP. The applicability of both Langmuir and Frendlich models suggest that there are both homogenous and heterogeneous active binding sites on the surfaces of RAOCP. These binding sites would be involved in monolayer as well as multilayer adsorption of the fluoride ions and the adsorption mechanism may, respectively, involve both chemisorption and physical adsorption processes (Bansal and Goyal, 2005; Hayes and Mabaga, 2013; Aziz and Mojin, 2014).

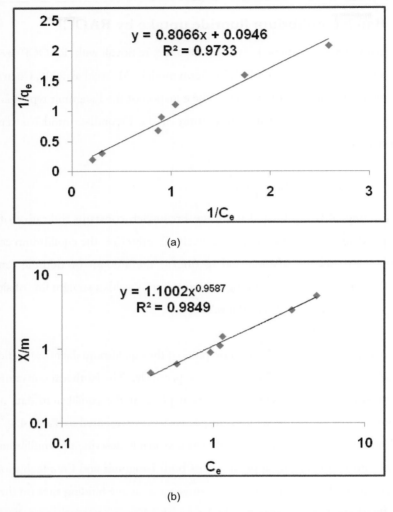

(a)

(b)

Fig. 4.7 (a) Langmuir and (b) Freundich isotherms for fluoride removal by RAOCP in batch adsorption experiment. Model water: fluoride = 5 ± 0.2 mg/L, HCO3 = 260 ± 5 mg/L, pH = 7.0 ± 0.1, adsorbent dose = 10 g/L, shaker speed = 100 rpm.

Table 4.4 attempts to provide a summarized comparison of fluoride removal performance of AOCP and RAOCP, including quantifications in terms of percentage improvements/increases as was found.

Chapter 4 Laboratory-scale column filter studies for fluoride removal with
aluminum (hydr) oxide coated pumice, regeneration and disposal

137

Table 4.4 Summarized comparison of fluoride uptake performance of RAOCP and AOCP under static equilibrium and dynamic adsorption conditions.

(A) Static equilibrium adsorption conditions

Adsorbent		Isotherm parameters and R^2 values				
		Langmuir			Freundlich	
	q_{max} (mg/g)	b (L/mg)	R^2	K_F (mg/g)x(mg/L)	n	R^2
AOCP	7.87	0.087	0.9853	1.1	1.043	0.984
RAOCP	10.57	0.1173	0.9733	-	-	-
% increase of capacity after regeneration	34 %	-	-	-	-	-

(B) *Dynamic fluoride adsorption conditions

Adsorbent	Fluoride adsorption capacity (q_b)(mg/g)	Service Time (h)	Volume of treated water (m³)	No. of Bed Volumes (BV)	Adsorbent usage rate (g/L)	Breakthrough model parameters							
						Thomas		Adams-Bohart			BDST		
						Q_{TH}	K_{TH} X10^{-2}	N_o	K_{AB} X10^{-2}	N_{bdst}	K_{ADS} X10^{-4}	Slope	Z_o
AOCP	2.32	66	0.9306	165	2.14	3.01	1.2	1120272.9	1.1	895.5	2.93	21.6	26.7
RAOCP	3.57	104	1.4664	259	1.44	5.16	0.5	1855955.0	0.5	1301.6	8.94	30.7	6.3
increase of efficacy of RAOCP	53.9%	57.6%	57.6%	57%	33% Less for RAOCP	71.4%	-	65.7%	-	43.34%	3 times more for RAOCP	1.4 times more for RAOCP	4.3 times more of AOCP

*Where applicable the comparison of dynamic performance was based on data for EBCT of 24mins, which was found to be optimum for the no. of bed volumes of water treated before breakthrough and the optimal use of adsorbent capacity for both AOCP and RAOCP

A comparison of the adsorption capacity of RAOCP with that of fresh AOCP under similar batch conditions applied in the previous studies, based on the Langmuir q_{max} (Table 3.6 of Chapter 3), revealed an increase of about 34 % after the regeneration, as indicated in Table 4.4, which suggests the effectiveness of the applied regeneration procedure.

4.5.2 Dynamic adsorption of fluoride onto RAOCP

The breakthrough concentration profiles at different bed depths for fluoride adsorption onto RAOCP are shown in Fig.4.8, which also follow the characteristic "S" shape, similar to that of AOCP.

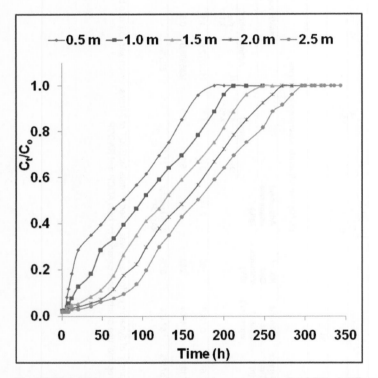

Fig 4.8 Breakthrough curves for fluoride removal by RAOCP at different bed depths: Filtration rate = 5.0 m/h; Model water: fluoride = 5.0 ± 0.1 mg/L, HCO_3^- = 330 ± 5 mg/L, pH = 7.0 ± 0.2, temp. = 20 °C

Chapter 4 Laboratory-scale column filter studies for fluoride removal with aluminum (hydr) oxide coated pumice, regeneration and disposal

139

A comparative plot of the breakthrough curves for RAOCP and AOCP is also presented in Fig. 4.9 (shown for two representative bed depths). The steepness of the breakthrough curves for RAOCP at all bed depths were found to be lesser than that of AOCP, which indicates a slower exhaustion of RAOCP compared to that of AOCP (Maliyekkal et al., 2006), and suggests an improved fluoride removal performance after the regeneration process.

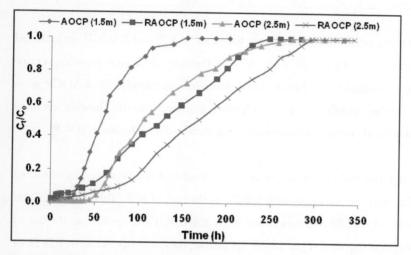

Fig 4.9. Comparison of breakthrough curves for fluoride removal onto RAOCP and AOCP for two bed depths. Filtration rate = 5.0 m/h; Model water: fluoride = 5.0 ± 0.1 mg/L, HCO_3^- = 330 ± 5 mg/L, pH = 7.0 ± 0.2, temp. = 20 °C

The service or breakthrough times (t_b) (h), fluoride adsorption capacities (q_b) (mg/g) (estimated from equation (4.1)), and treated water volumes (m^3)/number of bed volumes (BV) for RAOCP, presented in Table 4.5, showed a similar trend of variation with bed depth as that of AOCP.

Fluoride removal from groundwater by adsorption technology

Table 4.5 Bed depth, breakthrough times, volume of water treated and adsorption capacities for fluoride removal by RAOCP column filter

Bed depth (m)	EBCT (min.)	Mass of RAOCP (M_{RAOCP}) (g)	Volume of RAOCP (m³)	Breakthrough times (t_b) (h)	Volume of water treated (m³)	No. of bed volumes (BV)	RAOCP usage rate (g/L)	Fluoride adsorption capacity q_b (mg/g)
0.5	6	513	0.0014	20	0.282	200	1.82	2.743
1.0	12	1026	0.0028	48	0.677	239	1.52	3.292
1.5	18	1539	0.0042	76	1.072	253	1.44	3.475
2.0	24	2052	0.0057	104	1.466	259	1.40	3.567
2.5	30	2565	0.0071	120	1.692	239	1.52	3.292

A comparison of these performance indicators for fresh AOCP and RAOCP are included in Table 4.4. An increase of 54%, 58% and 57% of fluoride adsorption capacity, service time and volume of treated water before breakthrough, respectively, for RAOCP were observed which further confirmed a considerable improvement of fluoride removal performance under the dynamic conditions after regeneration of exhausted AOCP.

The adsorbent usage rate and bed life of the RAOCP fixed-bed system were also observed to follow a similar trend of variation with EBCT as that of AOCP (Tables 4.1, 4.5 and Fig.4.10). Similar to AOCP, 24 min was found to be the EBCT for optimizing the use of the RAOCP adsorption capacity. Thus where applicable, data corresponding to EBCT of 24 min were used in Table 4.4 for the summarized comparison of the dynamic performance of fluoride removal by AOCP before and after regeneration.

A reduction of about 33% of adsorbent usage rate for RAOCP and an increase of 57% for the bed life (Table 4.4) were observed, which were also indications of improved adsorption efficiency/performance after the regeneration.

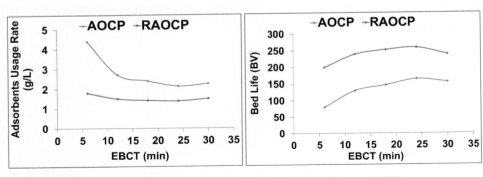

Fig. 4.10 (a) Adsorbent usage rate vrs EBCT   (b) Bed life vrs EBCT

Chapter 4 Laboratory-scale column filter studies for fluoride removal with
aluminum (hydr) oxide coated pumice, regeneration and disposal

141

Similar to fresh AOCP, the Thomas, Adams-Bohart and BDST models were found to be appropriate for describing the dynamic behavior of fluoride adsorption onto RAOCP. Comparative plots of the RAOCP experimental breakthrough curves, and that calculated using the non-linear forms of the Thomas and Adams-Bohart models are presented in Fig. A4.1 (a) and (b), respectively, in Annex 4, which show good correlations at all bed depths. The Thomas model constants (K_{TH} and q_{TH}) and those for the Adams-Bohart model (N_o and K_{AB}) for the RAOCP-fluoride system, including the χ^2 statistics are presented in Table A4.1 in Annex 4.The adsorption capacities determined from the Thomas model (q_{TH}) and Adams-Bohart model (N_o), were also found to be considerably higher for RAOCP compared to that of AOCP at all bed depths (Tables 4.1, A1. and 4.4).

The BDST plot for the RAOCP column experimental data is also presented on the same axial setting as that for AOCP (Fig 4.6), for ease of comparing their performances and the model parameters are included in Table 4.4

From Table 4.4 the adsorption capacity (N_{bdst}), determined from the BDST model was about 43% higher after the regeneration. The critical adsorbent bed depth Z_o (cm), required for producing a fluoride effluent concentration of acceptable quality (C_b of 1.5 mg/L) was 6.25 cm for RAOCP, which was about 4 times less than that of AOCP, suggesting a better fluoride removal potential of RAOCP. The obtained adsorption rate constant (K_{ads}) from the model for the fluoride removal in the RAOCP column bed was also 3 times more than that of fresh AOCP (Table 4.4), which signifies a greater likelihood of a delayed breakthrough for the fluoride-RAOCP system compared to that of the fluoride-AOCP system (Gorai and Pant, 2005). The slope of the BDST line (Fig.4.6 and Table 4.4) for the fluoride-RAOCP system was 30.7 (min/cm), hence about 1.4 times more than that for AOCP (= 21.6 min/cm). This indicates that under the same experimental conditions (filtration rate, influent concentration), the MTZ for the fluoride-RAOCP system would travel at a slower rate through the fixed-bed adsorbent column as compared to the MTZ for the fluoride-AOCP system (Taty-Costodes et al., 2005; Ghorai and Pant, 2005). This therefore supports the observed delayed breakthrough or increase of service time, and consequent improvement of about 57% in number of bed volumes of defluoridated water after the regeneration procedure.

A review of literature on regeneration of adsorbents reported by others, reveals that generally, the adsorption capacity/efficiency for most adsorbents are either fully (100%) restored or experiences a marginal loss (about 5 - 10%) or a considerable loss (up to 20%) of capacity after the first cycle regeneration (Bishop & Sansoucy, 1978; Younghun et al., 2004; Ghorai and Pant, 2004; Maliyekkai et al., 2006; Chauhan et al., 2007; Maliyekkai et al., 2008; Chen et al., 2011; Sun et al., 2011; Tresintsi et al., 2014; Qiusheng et al., 2014). In contrast, however, it was found in this study that the adsorption capacity of exhausted AOCP after the first cycle regeneration was not only fully (100%) restored, but attained a considerable increase: about 30% improvement of fluoride adsorption capacity under batch conditions, and generally more than 50% improvement under dynamic conditions, based on the assessed performance indicators (Table 4.4). This therefore suggests regeneration procedure explored in this study was effective and potentially useful.

As discussed in section 3.3.7 (chapter 3), created surface aluminol (Al–OH) functional groups onto the virgin pumice (VP) particles during the Al coating process for the synthesis of fresh AOCP, constitute the main active sites responsible for its characteristic reactions in the fluoride adsorption process. The superior fluoride removal performance of RAOCP therefore suggests that a higher quantity of the surface aminol (Al-OH) functional groups might have been created onto RAOCP during the regeneration process (Al re-coating), compared to that on fresh AOCP. This was most probably due to differences in the Al coating solution–base material interfacial properties, hence differences in the Al coating reaction/mechanism in the preparation of fresh AOCP, and the Al re-coating step for regeneration/restoration of the capacity of fluoride-saturated AOCP (FAOCP) (i.e. exhausted AOCP). The mechanism of the Al coating onto each of the base materials (i.e VP and FAOCP) is presumed to be predominantly a sorption process. XRF characterization indicates virgin pumice (VP) as a heterogeneous material, composed mainly of the oxides of Al, Si and Fe, that is known to possess cation exchange properties (Guler and Sarioglu, 2014; Tapan et al., 2013; Salifu et al., 2013; Sheet and Grayson, 1979). VP particle surfaces can exhibit positive charges at low pH environments, originating from

Chapter 4 Laboratory-scale column filter studies for fuoride removal with
aluminum (hydr) oxide coated pumice, regeneration and disposal

143

the protonation of hydroxyl functional groups (SiOH, AlOH, FeOH) in the Si, Al, and Fe bearing minerals in the pumice particles (Goldberg et al., 1999; Stumm and Morgan, 1996). The protonation reactions may result in the VP particle surfaces possessing cation exchange capacity, whereby the adsorbed protons (H^+) become the cation exchangeable sites. The sorption mechanism of Al species from the coating solution onto virgin pumice (VP) particles during the synthesis of AOCP, presumably involves predominantly a cation exchange process in which Al^{3+} ions replace the bound protons. Due to the higher valency of Al^{3+}, it would have a higher replaceability, and can consequently replace bound protons (H^+) from the VP particle surfaces, and be preferentially retained in accordance to the lyotropic series (Guler and Srioglu, 2014; Goldberg, 2010; Mohammad, 2004; GES, 2000; Wiese and Healy, 1974). On the other hand, with the applied regeneration approach whereby the adsorbed fluoride ions were not stripped of the surfaces of FAOCP prior to the regeneration (re-coating) process, the surface charges of the (FAOCP) particles may be influenced by the adsorbed fluoride ions, and would comparatively be negative (Goldberg, 2010). Consequently the sorption mechanism of Al^{3+} at the FAOCP-Al solution interface during the regeneration (i.e. Al re-coating) step may likely be dominantly a surface complexation process (Guler and Sarioglu, 2014; Figueroa and Mackay, 2004). Thus even though similar Al coating procedures were applied for the synthesis of fresh AOCP, as well as the restoration of the adsorption capacity of exhausted AOCP, the coating reactions/mechanisms were presumably different. In particular, the presence of adsorbed fluoride on FAOCP particle surfaces may enhance the Al^{3+} sorption process in the regeneration step, due to the high affinity of aluminum for fluoride, in accordance with the hard soft acid base (HSAB) concept (Alfara et al., 2004; Pearson, 1988), as presented in section 3.1 of the previous chapter. The average Al content of RAOCP (= 66 mgAl/g) extracted by acid digestion was found to be higher than that of fresh AOCP (= 22 mgAl/g) in chapter 3 (section 3.3.1). The higher Al content most probably resulted in the creation of a higher number of aluminol (Al-OH) functional groups onto RAOCP particle surfaces during the regeneration process, as compared to that on fresh AOCP, hence the higher fluoride removal performance of the former.

Based on the results of the first regeneration cycle which resulted in an increase of the fluoride adsorption capacity, and based on the application of the HSAB concept in the renegeration process as discussed, there is a good probability that further regeneration cycles could lead to further increase in the adsorption capacity. The expectation of the increase of the fluoride adsorption capacity in further regeneration cycles is illustrated schematically in Fig. 4.11. This, however, require to be verified through multiple regeneration experiments. The multiple regeneration experiments may help elucidate the trend of fluoride removal performance of RAOCP after each regeneration cycle, as this may possibly impact positively on the practical and economic viability of AOCP as a water defluridation adsorbent, if the trend remains positive. Further surface characterization, including the quantification of the aluminol adsorption sites of AOCP before and after regeneration is also required to further help in explaining the increase in fluoride removal performance of RAOCP.

Chapter 4 Laboratory-scale column filter studies for fluoride removal with aluminum (hydr) oxide coated pumice, regeneration and disposal

145

Fig. 4.11 Schematic illustration of the synthesis of fresh AOCP, regeneration concept/approach for RAOCP, presumed Al coating mechanisms/reactions and presumed percentage improvement of performance per regeneration cycle and increasing practical and economic viability per cycle, when the adsorbent is in use

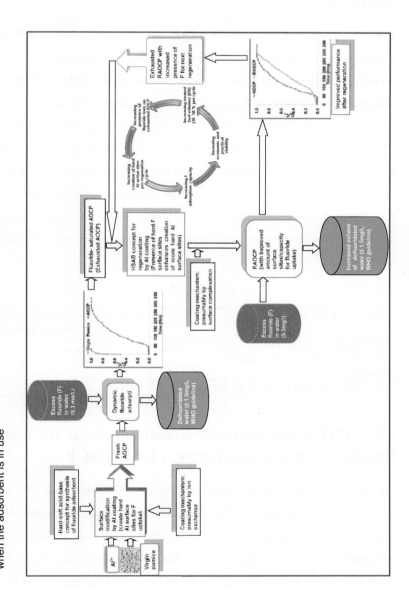

4.6   Leaching test for waste (spent) AOCP for safe disposal

Table 4.6 presents the results of TCLP extraction of spent AOCP (with and without stabilization). The fluoride concentrations of the extracts from both waste adsorbents were found to be less than the thresholds established by US-EPA, i.e 100 times the National Interim Primary Drinking Water (NIPDW) Standard for fluoride ((1.4 mg/L at 32° C). X 100 = 140 mg/L). None of the waste (spent) AOCP was thus hazardous according to the US-EPA criteria, and could be disposed off safely into a simple landfill. This suggests binding of the adsorbed fluoride ions on the spent AOCP surfaces was strong enough to prevent its mobility into the environment. Treatment of spent AOCP by coating with a final layer of Al for stabilizing fluoride, however, significantly reduced the mobility, as evident from the reduced fluoride concentration of the extract as shown in Table 4.6.

Table 4.6 TCLP extracts from spent AOCP.

Spent AOCP	Average fluoride concentration (mg/L) in triplicate samples
Untreated (without fluoride stabilization)	6.64 ± 0.15
Treated by Al coating (fluoride stabilized)	1.37 ± 0.02
NIPDWS regulatory level at 32°C	140

4.7   FTIR and thermodynamic analysis for insight into fluoride removal mechanism by RAOCP

Fig 4.11 shows the Fourier-transform infrared (FTIR) spectra of RAOCP (a) before and (b) after fluoride adsorption. The spectra of RAOCP before fluoride adsorption showed a broad adsorption band at 3000 - 3500 cm^{-1} which may be related to the stretching vibrations of O-H bonds (Kumar et al., 2011). Peaks were also observed at 969 cm^{-1} which corresponded to Al-OH mode, and at 690 cm^{-1} in the low absorption frequency, and both peaks suggested the presence of gibbsite on the RAOCP surfaces (Frost et al., 1999). After the adsorption of fluoride ions, the peak intensities related to the O-H stretching and Al-OH bending mode were reduced, which indicated the involvement of hydroxyl groups in the RAOCP-fluoride interactions.

Chapter 4 Laboratory-scale column filter studies for fuoride removal with aluminum (hydr) oxide coated pumice, regeneration and disposal

147

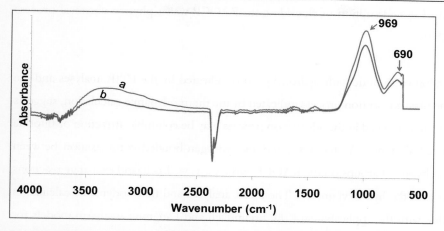

Fig 4.12 FTIR spectra of RAOCP (a) before and (b) after fluoride adsorption

Thermodynamic parameters such as the Gibbs free energy of adsorption (ΔG_{ads}) obtained from equilibrium isotherm data, can also provide an insight into the type and mechanism (physisorption or chemisorption) of the RAOCP-fluoride interaction. The Gibbs free energy of adsorption (ΔG_{ads}) was evaluated from the equation:

$$(\Delta G_{ads}) = -RT \ln(K) \tag{4.11},$$

where K is the equilibrium constant, related to the Langmuir isotherm constant, b (L/mole), based on equation (4.12), with the value 55.5 corresponding to the molar concentration of solvent (water), R is the molar gas constant (8.134 J/mole), and T is temperature in Kelven (Kumar et al., 2011).

$$K = 55.5 \times b \tag{4.12}$$

A negative ΔG_{ads} value (= -28.44 kJ/mole) obtained in this study indicates the fluoride-RAOCP interaction is spontaneous. Generally, the ΔG_{ads} value for physical adsorption is in the range of -20 kJ/mole ~ 0 and that for chemisorption is in the range -80 to -400 kJ/mole. Thus the other of magnitude of the ΔG_{ads} (= -28.44 kJ/mole) suggests the mechanism of the adsorption of fluoride onto RAOCP was a physico-chemical process (Yu et al., 2004).

The interaction of fluoride with hydroxyl groups indicated by the FTIR analyses and the physico-chemical adsorption process indicated by the thermodynamic calculation, suggests the mechanisms involved in the adsorption process may be columbic attraction of fluoride by protonated aluminol (AlOH) sites, and/or hydrogen-bonded complexation between fluoride and hydroxyl groups on the RAOCP surfaces, as well as ligand exchange reactions of fluoride with the hydroxyl groups. The FTIR analyses and thermodynamic calculation is also in line with the applicability of the Freundlich and Langmuir isotherm models for describing the adsorption equilibrium data for the fluoride-RAOCP system. The applicability of the two models suggested the adsorption mechanism is complex and involves both physical adsorption and chemical adsorption processes, and therefore supports/complements the thermodynamic and FTIR analysis.

4.8   Conclusions

ACOP was found capable of water defluoridation under continuous flow conditions, and an empty bed contact time (EBCT) of 24 min was found a suitable guide for design of full-scale water treatment systems, that would allow an optimal use of its fluoride adsorption capacity.

The Adam-Bohart model could predict very well the initial region of the of the breakthrough curves for the fluoride–RAOCP system, while the full breakthrough curve could be adequately descried by the Thomas and BDST models. The model parameters determined may be helpful for scaling up the process (for other flow rates and fluoride concentrations) for drinking water defluoridation using AOCP column filters, without need for additional experimentation, which could possibly save time and cost.

The fluoride adsorption capacity of exhausted AOCP, after the first cycle of regeneration was not only fully (100%) restored, but increased by more than 30% under batch conditions, and more than 50% under continuous flow conditions, suggesting the

Chapter 4 Laboratory-scale column filter studies for fuoride removal with
aluminum (hydr) oxide coated pumice, regeneration and disposal

149

effectiveness of the regeneration approach. This possible contribute to the economic viability of AOCP as water defluoridation adsorbent.

Presence of adsorbed fluoride on the surfaces of exhausted AOCP particles during regeneration process, presumably enhanced the creation of higher number of aluminol functional groups on RAOCP surfaces compared to that on fresh AOCP for fluoride adsorption and, resulted in a higher adsorption capacity of the former. Surface characterization, including quantification of the aluminol (Al-OH) adsorption sites of AOCP before and after regeneration is required to further help in explaining the increase of fluoride removal performance of RAOCP.

The effect of multiple regeneration cycles also need to be further investigated to assess the trend of fluoride removal performance of RAOCP, as this could positively impact on the practical and economic viability of AOCP as water defluridation adsorbent.

Waste (spent) AOCP (with or without stabilization/treatment) was found non-hazardous, and could be disposed of safely into a simple landfill without risk of environmental/groundwater re-contamination.

References

Aksu, Z., F. Gönen, F., 2004. Biosorption of phenol by immobilized activated sludge in a continuous packed bed: prediction of breakthrough curves, Process Biochemi. 39, 599–613.

Alfarra, A., Frackowiak, E.; Beguin, F., 2004. The HSAB concept as a means to interpret the adsorption of metal ions onto activated carbons. Applied Surf. Sci. 228, 84-92

Allen, S.J., Gan, Q., Mathew, R., Johnson, P.A., 2003. Comparison of optimized isotherm models for basic dye adsorption by kudzu. Bioresour Technol. 88, 143 -152.

Aziz, H.A., A. Mojiri, A., 2014. Wasewater Engineering: Advanced wastewater treatment systems. IJSR Publications, Penang, Malasia.

Apambire, W.B., Boyle, D.R., Michel, F.A., 1997. Geochemistry, genesis, and health implications of fluoriferous groundwaters in the upper regions of Ghana. Environ. Geol. 33 (1), 13–24.

Ayoob, S., Gupta, A.K., 2008. Insights into isotherm making in the sorptive removal of fluoride from drinking water. J. Hazard. Mater. 152, 976–985.

Bansal, R.C., Goyal, M., 2005. Activated Carbon Adsorption. CRC Press, Florida.

Bishop, P.L., Sansoucy, G., 1978. Fluoride removal from drinking water by fluidized activated alumina. J. Am. Water Works Assoc. 70, 554–559.

Chauhan, V.S., Dwivedi, P.K., Iyengar, L., 2007. Investigations on activated alumina based domestic defluoridation units J. Hazard. Materials. B139, 103–107.

Chen, N., Zhang, Z., Feng, C., Li, M., Chen, R., Sugiura, N., 2011. Investigations on the batch and fixed-bed column performance of fluoride adsorption by Kanuma mud. Desalination. 268, 76–82.

Chen, N., Zhang, Z., Feng, C., Zhu, D., Yang, Y., Sugiura, N., 2011. Preparation and characterization of porous granular ceramic containing dispersed aluminum and iron oxides as adsorbents for fluoride removal from   aqueous solution. J. Hazard. Mater. 186, 863–868.

Faust, S.D., Aly, O.M.   Chemistry of water treatment. 2nd ed, CRC Press LLC, Florida, USA, 1998.

Chapter 4 Laboratory-scale column filter studies for fluoride removal with
aluminum (hydr) oxide coated pumice, regeneration and disposal

151

Fawell, J., Bailey, K., Chilton, J., Dahi, E., Fewtrell, L., Magara, Y. Fluoride in drinking water. IWA Publishing, London.2006 4-81.

Figueroa, R.A., Mackay, A.A., 2004. Modeling Tetracycline Antibiotic Sorption to Clays Environ. Sci. Technol. 38, 476-483.

Frost, R.L., Kloprogge, J.T., Russel, S. C., Szetu, J.L., 1999. Vibrational Spectroscopy and Dehydroxylation of Aluminum (Oxo) hydroxides: Gibbsite. Applied Spectros. 53 (4), 423 -434.

García-Sánchez, J.J., Solache-Ríos, M., Martínez-Miranda, V., Solís Morelos, C., 2013. Removal of fluoride ions from drinking water and fluoride solutions by aluminum modified iron oxides in a column system. J. Colloid Interface Sci., 407, 410–415.

GES 166/266, Soil Chemistry. Winter, 2000. Lecture Supplement 3. Solid-Water interface; www. GES 166_266items_HO-3 Interface.doc.

Ghorai, S., Pant, K.K., 2004. Investigation on the column performance of fluoride adsorption by activated alumina, Chem. Eng. J., 98,165-173.

Ghorai, S., Pant, K.K., 2005. Equilibrium, kinetics and breakthrough studies for adsorption of fluoride on activated alumina, Sep. Purif. Technol. 2, 165–173.

Goldberg, S., 2010. Competitive adsorption of molybdenum in the presence of phosphorus or sulfur on gibbsite. Soil Sci., 175(3), 105-110.

Goldberg, S., Davis, J.A., Hem, J.D., 1999. The surface chemistry of aluminium oxide and hydroxides, in: G. Sposito (Ed.), The Environmental Chemistry of Aluminum, CRC Press, Florida.

Grchev, T., Cvetkovska, M., Stafilov, T., Schultze, J.W. 1991. Adsorption of polyacrylamide on gold and iron from acidic aqueous solutions.Electro. Acta 36, 1315-1323.

Guler U.A and Sarioglu, M., 2014. Removal of tetracycline from wastewater using pumice stone: equilibrium, kinetic and thermodynamic studies, J. of Env. Health Sci. & Eng., 12:79.

Han, R., Zou, L., Zhao,X., Xu,Y., Xu,F., Li,Y., YuWang., 2009. Characterization and properties of iron oxide-coated zeolite as adsorbent for removal of copper (II) from solution in fixed bed column. Chem. Eng. J. 149, 123–131.

Hayes, R.E., Mabaga, J.P., 2013. Introduction to Chemical Reactor Analysis, Second ed. CRC Press, Florida.

Hendricks, D. Fundamentals of water treatment unit processes: physical, chemical and biological, CRC Press, Taylor and Francis group, IWA Publishing, Florida, USA, 2011.

Ho, Y.S., 2004. Selection of optimum sorption isotherm. Carbon. 42, 2115 -211.

Kumar, E., Bhatnagara, A., Kumar, U., Sillanpaa, M., 2011. Defluoridation from aqueous solutions by nano-alumina: Characterization and sorption studies. J. of Hazard. Mater. 186, 1042–1049.

Kundu, S., Gupta, A.K., 2006. Arsenic adsorption onto iron oxide-coated cement (IOCC): regression analysis of equilibrium data with several isotherm models and their optimization, Chem. Eng. J. 122, 93–106.

Maliyekkal, S.M., Sharma, A.K., Philip, L., 2006. Manganese-oxide-coated alumina: a promising sorbent for defluoridation. Water Res. 40, 3497–3506.

Maliyekkal, S.M., Shukla, S., Philip, L., Nambi, I.M., 2008. Enhanced fluoride removal from drinking water by magnesi-amended activated alumina granules. Chem. Eng. J, 140, 183 -192.

Margui, E., Salvadó, V., I. Queralt, I., Hidalgo, M., 2004. Comparison of three-stage sequential extraction and toxicity characteristic leaching tests to evaluate metal mobility in mining wastes. Analytica Chimica Acta. 524, 151–159.

Mohammad, N., 2004. Hydraulics, diffusion and retention characteristic of inorganic chemicals in bentonite. PhD Dessertation, University of South Florida, Flotida, USA.

Negendra, C. R., 2003. Fluoride and Environment—A Review. Third international conference on environment and health, Chennai, India, 15th-17th December.

Pearson, R.G., 1988. Absolute electronegativity and hardness: Application to inorganic chemistry. Inorg. Chem. 27, 734-740.

Physick, A.J.W., Wales, D.J., Owens, S.H.R., Shang, J., Webley, P.A., Mays, T.J., Ting, V.P. 2016. Novel low energy hydrogen–deuterium isotope breakthrough separation using a trapdoor zeolite, Chem. Eng. J., 288, 161–168.

Qiusheng, Z.; Xiaoyan, L.; Bin, L.; Luo Xuegang, L., 2014. Fluoride adsorption from aqueous solution by aluminum alginate particles prepared via electrostatic spinning device. Chem. Eng. J. 256, 306 -315.

Quintelas,C., Pereira, R., Kaplan, E., Teresa Tavares, T., 2013. Removal of Ni(II) from aqueous solutions by an Arthrobacter viscosus biofilm supported on zeolite: From laboratory to pilot scale. Bioresour. Technol. 142, 368–374.

Rosenqvist, J., 2002. Surface Chemistry of Al and Si (hydr) oxide with Emphasis on nano-sized Gibbsite (Al(OH)3) Department of Chemistry, Inorganic Chemistry, Umea University, Sweden.

Salifu, A., Petrusevski, B., Ghebremichael, K., Modestus, L., Buamah, R., Aubry, C., Amy, G.L., 2013. Aluminum (hydr) oxide coated pumice for fluoride removal from drinking water: Synthesis, equilibrium, kinetics and mechanism. Chem. Eng. J. 228, 63 -74.

Salifu, A., Petrusevski, B., Ghebremichael, K., Buamah, R., G. Amy, G., 2012. Multivariate statistical analysis for fluoride occurrence in groundwater in the Northern region of Ghana J. Contaminant Hydro. 140–141, 34–44.

Sheet, P.D and Grayson, D.K., 1979. Volcanic Activity and Human Ecology. Academic Press, New York.

Shih, T.C., Wangpaichitr, M., Suffet, M., 2003. Evaluation of granular activated carbon technology for the removal of methyl tertiary butyl ether (MTBE) from drinking water. Water Res. 37, 375–385

Shomar, B., Muller, G., Yahya, A., Aska, S., Sansur, R., 2004. Fluoride in groundwater, soil and infused black tea and the occurrence of dental fluorosis among school children of the Gaza strip. J. Water and Health. 2 (1), 23–35.

Stumm, W and Morgan, J.J., 1996. Aquatic Chemistry Chemical Equilibria and Rates in Natural Waters. 3rd ed., John Wiley & Sons, Inc., New York, 1996, 1022 pp.

Sulaiman, A., Gupta, A.K., Basheer, A.B., 2009. A fixed bed sorption systems for defluoridation of groundwater. J. of Urban and Environ.Eng. 3, 17–22.

Sun, Y., Fang, Q., Dong, J., Cheng, X., Xu, J., 2011.Removal of fluoride from drinking water by natural stilbite zeolite modified with Fe(III). Desalination, 277, 121–127.

Tang, Y., Guan, X., Wang, J., Gao, N., McPhail, M.R., Chusuei, C.C., 2009. Fluoride adsorption onto granular ferric hydroxide: Effects of ionic strength, pH, surface loading, and major co-existing anions. J. of Hazard. Mater. 171, 774–779.

Chapter 4 Laboratory-scale column filter studies for fuoride removal with
aluminum (hydr) oxide coated pumice, regeneration and disposal

155

Tapan, M., Depci, T., Ozvan, A., Efe, T., Oyan, V., 2013. Effect of physical, chemical and electrokinetic properties of pumice on strength development of pumice blended cements, Mat. and Struc. 46, 1695 -1706.

Taty-Costodes, V.C., Fauduet, H., Porte, C., Hoa, Y., 2005. Removal of lead (II) ions from synthetic and real effluents using immobilized *Pinus sylvestris* sawdust: Adsorption on a fixed-bed column. J. Hazard. Mater. B123, 135–144.

Thomas, W.J., Crittenden, B. Adsorption Technology & Design. Reed Educational and Professional Publishing, Woburn, U.K, 1998.

Tresintsi, S., Simeonidis, K., Katsikini, M., Paloura, E.C., Bantsis, G., Mitrakas, M., 2014. A novel approach for arsenic adsorbents regeneration using MgO. J. of Hazard. Mater. 265, 217– 225.

US-EPA. 1992. Toxicity Characteristic Leaching Procedure (TCLP), U.S, EPA. Test methods for evaluating solid waste, Physical/Chemical methods, SW-846.

WHO, 2011. Guideline for Drinking Water Quality Incorporating. Forth edition. World Health Organization, Geneva.

Wiese, G.R and T.W. Healy, T.W., 1974. Adsorption of Al(III) at the TiO_2 -H_2O interface. J. of Colloid and Interf. Sci. 51(3), 434 - 442.

Younghun, K., Kim, C., Choi, I., Rengaraj, S., Yi, J., 2004. Arsenic Removal using Mesoporous Alumina prepared via a Templating Methods. Environ. Sci. and Technol. 38, 924-931.

Yu, Y., Zhuang, Y., Wang, Z., Qiu, M. 2004. Adsorption of water-soluble dyes onto modified resin. Chemosphere 54, 25–430.

Zheng, H., Hanb, L., Maa, H., Yan Zheng, Y., Zhang, H., Liu, D., Liang, S.,2008.
     Adsorption characteristics of ammonium ion by zeolite 13X. J. Hazard. Mater. 158,
     577–584.

Chapter 4 Laboratory-scale column filter studies for fluoride removal with
aluminum (hydr) oxide coated pumice, regeneration and disposal

157

Annex 4

Comparative plots for RAOCP-fluoride experimental breakthrough curves and calculated curves using the Thomas and Adams-Bohart models, related model constants and χ^2 statistics.

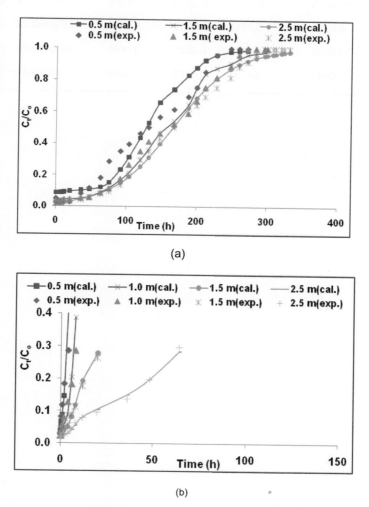

(a)

(b)

Fig. A4.1 Comparative plots of RAOCP experimental breakthrough curves, and that calculated using the non-linear forms of the (a) Thomas model and (b) Adams-Bohart model.

Table A4.1: Thomas and Adams-Bohart parameters at different bed depths for fluoride removal in the RAOCP fixed bed system.

Bed depth(m)	Thomas model			Adams-Bohart model		
	K_{TH}	Q_{TH}	X^2	K_{AB}	N_o	X^2
0.5	0.0064	10.1008	0.7628	0.0258	1352489.09	0.0831
1.0	0.0055	6.8557	0.2323	0.0104	1673786.14	0.0700
1.5	0.0051	5.8555	0.0852	0.0060	1995149.31	0.0483
2.0	0.0048	5.1579	0.0588	0.0005	1855955.02	0.0128
2.5	0.0046	4.6594	0.0469	0.0044	1783069.81	0.0050

Chapter 4 Laboratory-scale column filter studies for fuoride removal with aluminum (hydr) oxide coated pumice, regeneration and disposal

159

Leaching of aluminium and other selected metals under continuous flow conditions at 2.5 m bed depth with AOCP column filter.

AOCP		2.5 m Bed Depth (S5)															
		Unfiltered								Filtered							
		Al	Fe	Ca	K	Mg	Mn	Zn	Pb	Al	Fe	Ca	K	Mg	Mn	Zn	Pb
Concentration of Al and other selected metals in model water and filtrate from column filter (mg/L)	Model water after 0.5 Hrs	0.014	<0.004	50.200	5.500	6.260	<0.004	0.007	<0.004	0.041	<0.004	49.540	6.020	6.220	<0.004	0.017	<0.004
	Effluent after 0.5 Hrs	0.051	<0.004	13.260	6.800	3.820	0.050	0.009	<0.004	0.081	<0.004	13.700	7.480	3.860	0.048	0.014	<0.004
	Model water after 78 Hrs	0.024	<0.004	49.620	5.440	6.200	<0.004	0.005	<0.004	0.045	<0.004	49.580	5.680	6.300	<0.004	0.009	<0.004
	Effluent after 78 Hrs	0.033	<0.004	19.460	5.400	6.040	<0.004	<0.004	<0.004	0.044	<0.004	19.580	5.640	6.020	<0.004	<0.004	<0.004
	Model water after 264 Hrs	0.019	<0.004	47.540	5.360	6.260	<0.004	0.072	<0.004	0.068	<0.004	47.600	5.480	6.240	<0.004	0.008	<0.004
	Effluent after 264 Hrs	0.144	<0.004	27.918	5.440	6.180	<0.004	<0.004	<0.004	0.136	<0.004	27.140	5.580	6.100	<0.004	0.004	<0.004
Net Concentration due to use of AOCP (mg/L)	Effluent after 0.5 Hrs	0.037	<0.004	0	1.300	0	0.046	0.001	<0.004	0.040	<0.004	0	1.460	0	0.044	0	<0.004
	Effluent after 78 Hrs	0.008	<0.004	0	0	0	<0.004	0	<0.004	-0.001	<0.004	0	0	0	<0.004	0	<0.004
	Effluent after 264 Hrs	0.125	<0.004	0	0.080	0	<0.004	0	<0.004	0.068	<0.004	0	0.100	0	<0.004	0	<0.004
WHO Guideline Value (mg/L)		0.200	0.300	–	–	–	0.400	–	0.010	0.200	0.300	–	–	–	0.400	–	0.010
Remarks		Ok!	Ok!	Ok!	Ok!	Ok!	Ok!	Ok!	Ok!	Ok!	Ok!	Ok!	Ok!	Ok!	Ok!	Ok!	Ok!

Leaching of aluminium and other selected metals under continuous flow conditions at 2.5 m bed depth with RAOCP column filter

RAOCP		Unfiltered								Filtered (2.5 m Bed Depth (85))							
		Al	Fe	Ca	K	Mg	Mn	Zn	Pb	Al	Fe	Ca	K	Mg	Mn	Zn	Pb
Concentration of Al and other selected metals in model water and filtrate from column filter (mg/L)	Model water after 0.5 Hrs	<0.004	0.005	45.800	5.280	6.400	<0.004	0.008	<0.004	0.006	<0.004	42.580	5.020	5.940	<0.004	0.010	<0.004
	Effluent after 0.5 Hrs	0.028	<0.004	14.640	1.060	4.340	0.163	0.010	<0.004	0.008	<0.004	14.520	1.278	4.380	0.156	0.013	<0.004
	Model water after 80 Hrs	0.004	<0.004	43.620	5.060	6.140	<0.004	0.008	<0.004	0.010	<0.004	41.920	5.020	5.900	<0.004	0.010	<0.004
	Effluent after 80 Hrs	0.010	<0.004	42.060	5.020	6.120	<0.004	<0.004	<0.004	0.024	<0.004	41.280	5.060	5.960	<0.004	0.005	<0.004
	Model water after 264 Hrs	<0.004	0.004	44.360	4.960	6.140	<0.004	0.009	<0.004	<0.004	<0.004	44.420	5.260	6.180	<0.004	0.013	<0.004
	Effluent after 264 Hrs	0.088	<0.004	44.200	5.020	6.200	<0.004	<0.004	<0.004	0.078	<0.004	43.260	5.160	6.040	<0.004	0.005	<0.004
Net Concentration due to use of RAOCP (mg/L)	Effluent after 0.5 Hrs	0.024	-0.001	0	0	0	0.159	0.001	<0.004	0.001	<0.004	0	0	0	0.152	0.004	<0.004
	Effluent after 80 Hrs	0.006	<0.004	0	0	0	<0.004	0	<0.004	0.015	<0.004	0	0.040	0.060	<0.004	2E-05	<0.004
	Effluent after 264 Hrs	0.084	<0.004	0	0.060	0.060	<0.004	0	<0.004	0.074	<0.004	0	0	0	<0.004	3E-05	<0.004
WHO Guideline Value (mg/L)		0.200	0.300	–	–	–	0.400	–	0.010	0.200	0.300	–	–	–	0.400	–	0.010
Remarks		OK	OK	OK	OK	OK	OK	OK	OK	OK	OK	OK	OK	OK	OK	OK	OK

5

Fluoride removal from drinking water using granular aluminum-coated bauxite as adsorbent: Optimization of synthesis process conditions and equilibrium study

Main part of this chapter was published as:
Salifu, A., Petrusevski, B., Mwampashi, E.S., Pazi, I.A., Ghebremichael, K., Buamah, R., Aubry, C., Amy, G.L., Kenedy, M. D. 2016. Defluoridation of groundwater using aluminum-coated bauxite: Optimization of synthesis process conditions and equilibrium study. J. of Env. Manag. 181,108 -117.

Abstract

The section of the study reported in this chapter, explored the possibility of modifying the physico-chemical properties of bauxite, a locally available material in many countries including Ghana, by thermal treatment and an aluminum coating, for water defluoridation. The synthesis of granular aluminum-coated bauxite (GACB) as a fluoride adsorbent is described. The study mainly focused on investigating the effects of varying synthesis process conditions on the defluoridation efficiency of GACB, using series of batch adsorption experiments. GACB performed better than the raw bauxite (RB) base material, and was able to reduce fluoride concentration in model from 5 ± 0.2 mg/L to ≤ 1.5 mg/L (World Health Organization (WHO) guideline). Based on nonlinear Chi-square (χ^2) analysis, the best-fitting isotherm model for the fluoride-GACB system was in the order: Freundlich > Redlich-Perterson $\approx$ Langmuir > Temkin. The fluoride adsorption capacity of GACB ($q_{max} = 12.29$ mg/g) based on the Langmuir model, was found to be either comparable or higher than the capacities of some reported fluoride adsorbents. XRD characterization suggested aluminum was incorporated in GACB in the form of gibbsite by the applied coating process. The results also suggested that the gibbsite incorporated by the Al coating procedure, was more effective for fluoride uptake compared to the intrinsic (natural) gibbsite content of the raw bauxite base material. Presumably the crystal structure of gibbsite incorporated by the Al coating possesses a higher percentage of specific surface area of edge faces, hence more reactive sites for anion adsorption, than the intrinsic gibbsite. Thermal pre-treatment of raw bauxite (RB) prior to the Al coating, contributed significantly to an increase of the fluoride adsorption efficiency of the produced GACB. The applied procedures in this study could therefore be a useful approach for synthesizing an effective fluoride adsorbent using bauxite, a locally available material. Kinetic and isotherm analysis as well as FTIR and Raman analysis suggested the mechanism of fluoride adsorption onto GACB was complex, and involved both physical adsorption and chemisorption processes.

Chapter 5 Fluoride removal from drinking water using granular aluminum-coated bauxite as adsorbent: Optimization of synthesis process and equilibrium study

163

5.1   Background

Raw bauxite (RB) is a local material available in some developing countries including Ghana that could be used as a raw base material for synthesizing a fluoride adsorbent, especially in places where it may either be more readily available than pumice or where pumice may not be available. Raw bauxite (RB) has known defluoridation properties (Sujana and Anand, 2011). Moreover bauxite is robust, possess sufficient mechanical strength and could overcome limitations such as clogging and/or low hydraulic conductivities in fixed-bed adsorption systems. Fluoride adsorption capacity of raw bauxite is, however, limited and modifications of the physico-chemical properties for improved performance, appear not to have been well studied. This makes bauxite a good candidate as base material for surface modification studies in order to produce a fluoride adsorbent with enhanced uptake capabilities.

In chapter 3, the synthesis of a fluoride adsorbent by an aluminum coating (i.e. AOCP) and its efficacy for drinking water defluoridation was presented. Studies related to optimization of the synthesis process conditions for the production of fluoride adsorbent, aimed at opmtizing the efficacy has, however, not been conducted.

Because adsorption is a surface phenomenon, the rate and extent of adsorption could be influenced by the specific surface area of the adsorbent. The larger the specific surface area of an adsorbent, the greater the adsorption capacity could be expected (Armenante, 2009; Black & Veatch, 2007; U.S Army Engineers, 2001; Sawyer and McCarty, 1994). The properties which may therefore contribute to the number of available sites for fluoride adsorption by a modified bauxite may include both the textural properties (i.e surface area and pore voulme), as well as the nature of the surface (i.e. its reactivity/affinity for fluoride ions) (Cater et al, 1986).

Fluoride contaminated groundwater always contain other co-existing ions which can either enhance or compete and interfere negatively with the adsorption process (Kamble et al., 2007). The useful life time of an adsorbent may reduce in natural fluoride-cntaminated groundwater that contain high concentration of compiting co-ions, and therefore the effect of co-existing ions requires to be studied.

The main goal of this part of the study was to further explore the hard soft acid base (HSAB) concept, in combination with thermal pre-treatments of raw bauxite (RB), for optimizing the surface reactivity as well as enhancing the textural properties of the aluminium oxide modified bauxite for improved fluoride removal capabilities. A working hypothesis was that, a higher amount of Al coating would result in higher amount of hard surface sites, hence a corresponding increase in fluoride binding capability in accordance with the HSAB concept.

The specific objectives were to: (i) explore the possibility of synthesizing granular aluminum coated bauxite (GACB) as a fluoride adsorbent, with a focus on investigating the influence of varying synthesis process conditions on the amount of aluminum coating onto GACB as well as its surface area, (ii) study the fluoride adsorption potential of the synthesized adsorbent, and (iii) examine the effects of major co-anions on fluoride adsorption by GACB, as a preliminary assessment of its performance in natural groundwater.

Series of batch adsorption experiments were conducted for evaluating the fluoride removal efficiency of GACB produced under different process conditions, namely; varying coating pH and varying process temperatures for thermal pre-treatment of raw bauxite (RB) prior to surface modification by Al coating, aimed at optimizing the efficiency of GACB. FTIR and Raman spectroscopic analysis were employed to gain an insight into the mechanisms involved in the adsorption of fluoride onto GACB. These characterization techniques are helpful for studing the nature of the adsorbent material and the possible types of surface functional groups present, and how they interact with the fluoride during the adsorption process.

5.1.1 Surface reactivity of RB particles and potential for binding cations in aqueous solutions

Bauxite is a naturally occurring heterogeneous material composed primarily of one or more aluminum hydroxide/oxides minerals and other constituents including iron oxides, silica, titanium oxides, aluminosilicates and water (Gow and Lozej, 1993). By its composition, pH-dependent variabl-charged surfaces may exist on it when in aqueous solutions, along-side permanent negatively charged-surfaces. The origin of variable charge is the result of unsatisfied bonds at the terminal ends of the metal (M) oxides in the bauxite mineral assemblage, which leads to a dissociative chemisorption of water in aqueous solutions and result in the formation of surface hydroxyl functional groups (MOH) or active surface sites (Rosenqvist, 2002; GES 166/266, 2000). The surface functional groups may undergo protonation-deprotonation reactions which result in surface charges, the sign and magnitute of which is a function of the solution pH. The permanent (i.e. pH in-dependent) negatively charged surfaces stems partially from the presence of constituents such as aluminosilicates minerals. The negative charge arises from structural charge imbalance caused by isomorphic substitution within these minerals, whereby some cations are replaced by other cations of lower valency. Most commonly, Mg^{2+} replaces some Al^{3+} in octahedral sheets, or some Si^{4+} in tetrahedral sheets can be replaced by Al^{3+}, which may result in residual net negative charges on the bauxite particle surfaces (Grafe et al., 2009; Mohammad, 2004). The charge development on the bauxite particle surfaces would make it reactive and can consequently be involved in sorption-desorption reactions of ions at a bauxite mineral-water interface, e.g, during the Al coating process. The charged state (i.e charge density and sign) of the bauxite mineral surfaces may also be a significant regulator for aqueous ionic species being either removed from solution onto the bauxite surfaces or desorbed into solution (Grafe et al., 2009).

5.1.2 Hydrolysis of aluminuim

Aluminum metal ions readily undergo hydration reactions in aqueous systems. When an aluminium-bearing salt (eg. $Al_2(SO4)_3$ or $AlCl_3$) is added to water at concentrations lower than the solubility product constant of the amorphous hydroxide, it immediately dissociates, and the aluminium metal ions form coordination compounds with water molecules. The hydrolysis reactions of aluminium result in the formation of monomeric, dimeric and polymeric hydroxometal complexes. Aluminium is amphoteric and hence form both cationic and anionic complex species in aqueous solution (Faust and Aly, 1998; Bratby, 1980). Amorphous form of aluminium hydroxide precipitates whenever the solubility product for $Al(OH)_3(s)$ is exceded. Some of the hydrolysis reactions of aluminium (i.e. eqns 5.1 to 5.10) and the related equilibrium constant are given in Table 5.1 (Faust and Aly, 1998).

Table 5.1 Hydrolysis of aluminium and equilibrium constants.

Reaction	Log K_{aq}	Eqn
$Al^{3+} + H_2O = AlOH^{2+} + H^+$	-4.97	5.1
$Al^{3+} + 2H_2O = Al(OH)_2^+ + 2H^+$	-9.3	5.2
$Al^{3+} + 3H_2O = Al(OH)_{3(aq)} + 3H^+$	-15.0	5.3
$Al^{3+} + 4H_2O = AlOH_4^- + 4H^+$	-23.0	5.4
$2Al^{3+} + 2H_2O = Al_2(OH)_2^{4+} + 2H^+$	-7.7	5.5
$3Al^{3+} + 4H_2O = Al_3(OH)_4^{5+} + 4H^+$	-13.9	5.6
$13Al^{3+} + 28\ H_2O = Al_{13}O_4(OH)_{24}^{7+} + 32H^+$	-98.7	5.7
$\alpha\text{-}Al(OH)_{3(s)} + 3H^+ = Al^{3+} + 3\ H_2O$	8.5	5.8
$Al(OH)^3 + 3H^+ = Al^{3+} + 3\ H_2O$ amorph	10.5	5.9
$Al^{3+} + 3OH^- = Al(OH)^3(s)$	33.0	5.10

The aqueous speciation of aluminium changes as function of pH, concentration and counter ion (e.g. Cl^-, SO_4^{2-}), with the pH having the most remarkable effect among these parameters (Sarpola, 2007). Fig. 5.1 shows the effect of pH on the aqueous speciation and solubility of aluminium at equlibrium with (a) gibbsite and (b) amorphous $Al(OH)_3$, when $Al_2(SO4)_3$ is dissolved in water. This gives an indication of the aluminium species that may be present during the coating process for the synthesis of Al modified-based fluoride adsorbents, when using aluminium sulphate as the Al-bearing solution.

Chapter 5 Fluoride removal from drinking water using granular aluminum-coated bauxite as adsorbent: Optimization of synthesis process and equilibrium study

167

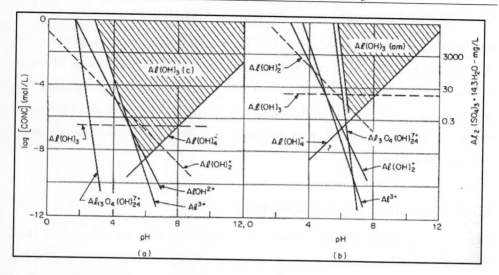

Fig. 5.1: Effect of pH on aqueous speciation and solubility of aluminium at equilibrium with (a) gibbsite and (b) amorphous Al(OH)$_3$ (Source: Faust and Aly, 1998).

5.2 Material and Methods

5.2.1 Synthesis of granular aluminum coated bauxite (GACB)

5.2.1.1 Coating Al onto bauxite particle surface at varying pH

RB used as base material in Al coating experiments for surface modification, was obtained from Awaso Bauxite Mines (Ghana). Bulk samples were crushed and sieved to a particle size range of 0.8-1.12 mm, thoroughly washed and air-dried for use in the Al coating process. Coating of RB was done by completely soaking about 150 g of dried sample in sufficient amount of 0.5M Al$_2$(SO$_4$)$_3$ in a Pyrex glass beaker, and followed a similar procedures as described for the coating of pumice (section 3.2.1.1). To investigate the effect of coating pH, however, Al coating was done at pH values of 1, 2, 4 and 6. NaOH and/or HCl was slowly added to the completely mixed RB – Al mixture for pH adjustment to the desired coating pH, using a small capacity peristaltic pump (Waston Marlow, 1014, U.K). The coated samples were well sieved and washed with demi-water buffered at pH 7.0 ± 0.1, to remove any loosely bound aluminum (hydr)oxide precipitates. This was to ensure only the effectively coated RB was used for subsequent fluoride adsorption studies to ensure reliability of the performance results.

5.2.1.2 Thermal pre-treatment of RB prior to Al coating

Samples of dried RB were pre-treated by calcination at different process temperatures (200, 400, 500, 600 and 700 °C) in a muffle furnace for 2 h, before being used in the Al coating process. The aim was to improve on the textural proprieties (i.e. surface area, pore size, pore volume) as well as enhance the amount of Al coating, hence nature of the coated bauxite surface. The combined effect was expected to result in increased fluoride uptake by the synthesized material.

5.2.2 Characterization of adsorbent

The physico-chemical properties of RB and GACB were determined based on BET, SEM/EDS and XRF characterization using similar techniques as described for AOCP in chapter 3. Aluminum contents of RB and GACB were also extracted by acid digestion as described for AOCP. In addition, Fourier transmission infra-red (FTIR) spectra of the adsorbents was collected using Nicolet 8700 FTIR spectrometer (Thermo instruments, USA). Raman spectra of RB, GACB and GACB with adsorbed fluoride were also obtained using inVia Raman microscope (Renishaw, U.K) and the Raman shifts were measured from 200 to 3,700 cm^{-1}

5.2.3 Batch adsorption experiments

Fluoride adsorption behaviors of GACB produced under different conditions and that of raw bauxite (RB) were studied in batch adsorption experiments at room temperature (20^0C) and neutral pH (7 $\pm$ 0.2) using similar methods as described for AOCP in chapter 3, including the analytical technique for determining the residual fluoride concentration in the filtrate sampled at pre-determined times during the experiments.

5.3 Results and discussion

5.3.1 Characterization

It was observed from SEM images of uncoated RB and GACB (Fig. 5.2a) that bauxite particle surfaces were covered with a layer of Al, which may be attributed to the surface modification by the coating process. Fig. 5.2b presents XRD spectrum of RB and GACB,

Chapter 5 Fluoride removal from drinking water using granular aluminum-coated bauxite as adsorbent: Optimization of synthesis process and equilibrium study

169

with major peaks labeled. The XRD patterns show that gibbsite $(Al(OH)_3)$ and hematite (Fe_2O_3) were the main components of RB. In comparison, XRD patterns obtained with GACB show a decrease in hematite peaks. This change in intensity suggest a possible incorporation of aluminum into GACB in the form of gibbsite phase. FTIR characterization of GACB (a) before and (b) after fluoride adsorption (Fig. 5.2c) was also employed to get an insight into the nature of GACB as well as its interaction with fluoride. GACB showed FTIR adsorption band at 3300 - 3700 cm-1, which may be assigned to the -OH stretching frequency of gibbsite. Bands also observed at 550 -1,200 cm-1 show the -OH bending deformation in GACB (Baral et al., 2007). The FTIR intensities of the -OH groups were, however, found to reduce after the fluoride adsorption indicating surface OH groups were involved in the interaction between fluoride and GACB surfaces (Okamofo & Imanaka, 1988). The Raman spectra of RB, GACB and after fluoride adsorption onto GACB is also shown in Fig. 5.2d. The spectrum of RB shows four bands at 3619, 3525, 3435 and 3365 cm-1 for gibbsite at the hydroxyl stretching region, and bands at 1425 and 1320 cm-1, indicating the presence of maghetite and hematite, respectively, in RB (Oh et al., 1998; Ruan et al., 2001) New weak bands were observed in the spectrum of GACB, including bands at 306 cm-1 for gibbsite and 414 cm-1 for aluminum oxide (Oh et al., 1998; Thoma et al., 1989), which may be attributed to the Al coating process. After the adsorption of fluoride onto the GACB surface, the Raman intensities were modified, and weak bands were observed at 322 and 546 cm-1, that may be related to Al - F interactions (Boulard et al., 1989; Gilbert et al., 1988).

The interaction of fluoride with hydroxyl ions indicated by the FTIR analysis and the presence of Raman bands indicating Al - F interactions, suggest that ligand exchange between F and hydroxyl groups on GACB surface may be involved in the adsorption mechanism.

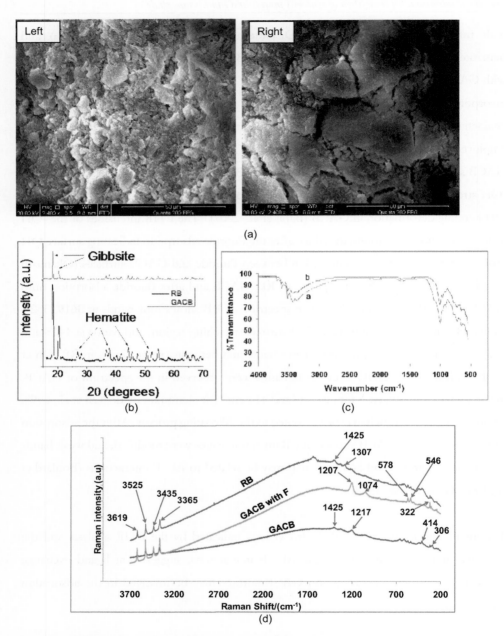

Fig. 5.2: (a) SEM micrographs of RB (left) and GACB (right), (b) XRD spectrum of RB (black) and GACB (red) (Curves are vertically shifted for more clarity), (c) FTIR spectra of GACB (a) before and (b) after fluoride adsorption and (d) Raman spectrum of RB, GACB and GACB with fluoride adsorbed.

Chapter 5 Fluoride removal from drinking water using granular aluminum-coated bauxite as adsorbent: Optimization of synthesis process and equilibrium study

171

Elemental composition (% by wt.) of RB and GACB coated at pH 2, obtained from XRF and EDX analyses are presented in Table 5.2. An increase of Al n GACB compared to RB, as well as emergence of sulfur in GACB were observed, which may also be attributable to the Al coating with $Al_2(SO_4)_3$.

Table 5.2 Elemental composition of RB and GACB coated at pH 2, obtained by XRF and EDX analyses.

Element			Al	Fe	Zr	Ti	Ru	Ca	Si	S	O
XRF (% by wt)		RB	74.1	8.2	5.0	2.8	1.2	1.2	1.7	0.31	-
		GACB	77.1	7.9	0.3	2.9	0.9	0.9	1.7	1.9	-
EDX (% by wt.)	with oxygen	RB	24.2	1.6	-	1	-	-	0.2	0	62.1
		GACB	27	0.4	-	0.1	-	-	0	0.8	63.4
	without oxygen	RB	88.6	5.9	-	3.7	-	-	0.7	0	-
		GACB	95.4	1.4	-	0.4	-	-	0	2.8	-

The N_2 adsorption-desorption isotherms for GACB produced from RB with no pre-treatment and that produced from thermally pre-treated RB (Fig. 5.3a) show a hysteresis loop which indicate GACB as a typical mesoporous material (Groen et al., 2003). The pore size distribution (Fig. 5.3b) also showed GACB as predominantly mesoporous (2 nm to 50 nm), which makes it an appropriate material for the uptake of fluoride ions (with radius of 1.33 Å) into its inner layers (Groen et al., 2003; Kaneko, 1994). The textural parameters of GACB calculated from the N_2 adsorption isotherms are included in Table 5.3 in the section where the "Effect of thermal pre-treatment of RB" is further discussed.

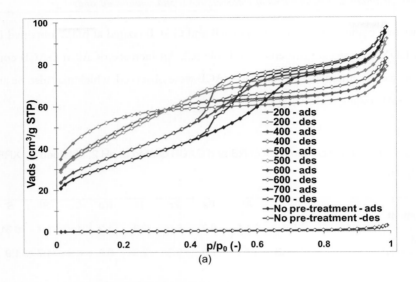

(a)

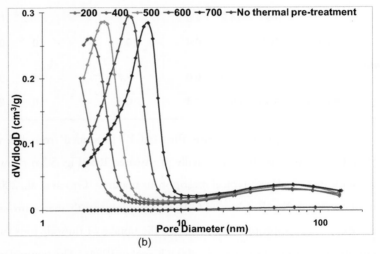

(b)

Fig. 5.3 (a) N_2 Adsorption-desorption isotherms for GACB and (b) Pore size distribution for GACB

5.3.2 Effect of coating pH and thermal pre-treatment on aluminum coating and fluoride removal efficiency

5.3.2.1 pH and net Al coating

Sorption of Al onto RB particles occurred over the entire coating pH range of 1 to 6, as shown in Fig. 5.4a. The net amount Al coated onto RB ranged between 27.4 and 164 mg Al/g, indicating the effectiveness of the coating procedure.

The mechanism of sorption of aqueous Al species onto the RB particle surfaces at the RB-water interface during the coating process, presumably involves ion exchange, in which Al^{3+} ions replace bound protons and/or electrostatic attraction onto the permanent negatively charged surfaces in the RB mineral assemblage. Due to the higher valance of Al^{3+}, it would have a higher replaceability, and, may therefore replace bound protons from the RB surfaces and be preferentially retained in accordance with the lytropic series (strength of retention) (Mohammad, 2004; GES, 2000; Wiese and Healy, 1994). The possible surface complexation of sulphate ions (in $Al_2(SO_4)_3$) onto RB surfaces may also contribute to its negative charges, which may also promote binding of Al ions (Goldberg et al., 2010).

The net amount of Al coated onto RB was found to be influenced by the coating pH. The coating efficiency increased with decreasing pH from 6 to 2, was most effective at coating pH 2, and thereafter declined with further decrease of pH (Fig. 5.4a). This may be due to the increasing Al solubility with decreasing pH, and hence increasing concentration of Al species in coating solution, a factor which may play an important role in the proton replacement from RB surfaces, and the preferential retention of Al ions (Mohammad, 2004). The increasing retention of Al species with decreasing pH from 6 to 2 may, however, induce an increasing positive charge on the RB particle surfaces. The pH where the Al coating start decreasing (pH 2) most likely represent the point where the RB surface charge is sufficiently positive such that a repel of the positively-charged Al ion species predominates, and rather promote its desorption with further lowering of the coating pH (Yan-Pang et al., 2013).

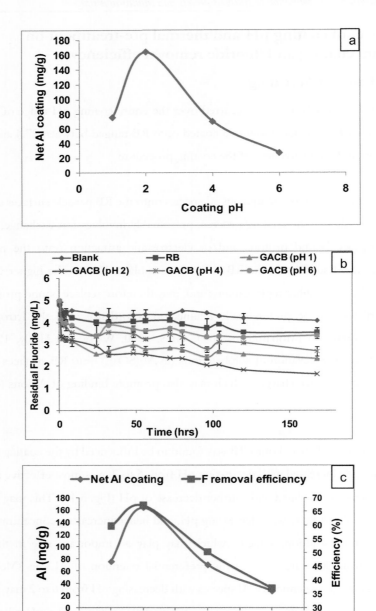

Fig. 5.4 (a) Net Al coating onto RB at different coating pH, (b) fluoride uptake in batch experiments by GACB coated under different pH conditions, and (c) comparison of trends of pH effect on net Al coating and fluoride uptake efficiency.

Chapter 5 Fluoride removal from drinking water using granular aluminum-coated bauxite as adsorbent: Optimization of synthesis process and equilibrium study

175

5.3.2.2 Fluoride scavenging potential of GACB

Fig. 5.4b presents results of batch adsorption experiments for assessing the fluoride removal potential of RB and GACB coated at different pH.

The fluoride removal efficiency of RB was found to be limited to 26.4% while that of GACB was up to 67.3%, indicating GACB possessed significantly higher fluoride removal capability, which may be attributable to the modifications of the RB particle surfaces by the Al coating. It was observed from the XRD analysis that the aluminum coating process resulted in an increase in the gibbsite content of GACB compared to hematite as originally found in RB. This suggests that the increased fluoride removal efficiency of GACB may be largely due to the increased content of gibbsite, an aluminum-based oxide, which has been observed by others to possess higher defluoridation capabilities than hematite, an iron-based oxide (Laveccehia, 2012). This is presumably due to differences in affinities between fluoride and the cations present in the two respective hydr(oxides), which complexes with hydroxyl groups that gives rise to the functional groups largely responsible for the fluoride adsorption. Al^{3+} (hardness (η) =45.77) present in gibbsite ($Al(OH)_3$) may posses a higher affinity for fluoride compared to Fe^{3+} (hardness (η) = 12.08) in hematite in accordance to the HSAB concept (Pearson, 1998; Pearson, 1993).

Similar to the net Al coating, the fluoride removal efficiency of GACB was found to be coating pH-dependent and also found to follow a similar trend with a change of coating pH. The fluoride removal efficiency increased with increasing net Al coating (hence increasing incorporation of gibbsite) and vice versa (Fig. 5.4b and 5.4c). The obtained results were thus in agreement with the working approach for synthesizing a fluoride adsorbent in this study. GACB synthesized at coating pH 2 exhibited the maximum fluoride scavenging potential, and was considered the optimum coating pH, and was employed for further work.

Further observation also suggested that the Al incorporated in the form of gibbsite by the applied coating procedure, was more effective for fluoride uptake compared to the intrinsic gibbsite content of RB. For instance, using the % by weight of Al as obtained from the EDX analysis (with oxygen) (Table 5.2) as measures/indicators of their respective gibbsite contents, the intrinsic (natural) Al content of RB was 24.2 % while that of GACB coated at pH 2 was 27 %, representing an increase of 11.6 % from the Al coating process. In contrast, however, the fluoride removal efficiency of RB at equilibrium was 26.4 %, whereas that of GACB coated at pH 2 was 67.3 %, representing an increase of about 155 % of the fluoride removal efficiency, i.e., about 13 fold increase in relation to the increase in gibbsite content. This may be due to differences in the reactivities of the gibbsite incorporated by the Al coating process and that of the natural gibbsite in RB. Gibbsite has been observed to crystallize into heterogeneous faces whereby, two types of crystal faces (i.e. edge and planner), with different reactivities towards anions can be distinguished. The planner surface comprises structural OH^- groups coordinated to two Al^{3+} ions, which is thought, makes the AlOH at the planner faces non-reactive/inert, due to the stability of the bonding. On the other hand, two types of OH^- groups are found at the edge faces, that is the doubly coordinated OH^- groups and also OH^- groups that are coordinated to single Al^{3+} ions. The singly coordinated OH^- groups at the edge faces are believed to readily protonate to form reactive $AlOH^+_2$ groups, which acts as Lewis acids sites for anion adsorption (Kumar et al., 2014; Jodin et al., 2005; Goldberg et al., 1996; Dobias et al., 1993; McBride and Wesselink, 1988). The present observation thus suggest the crystal structure of gibbsite incorporated by the Al coating procedure, presumably possess higher percentage of specific surface area of edge faces, hence more reactive sites for fluoride adsorption than the intrinsic gibbsite. The coating procedure could therefore be a useful approach for enhancing the fluoride uptake properties of bauxite and/or similar locally available materials for water defluoridation.

5.3.2.3 Effect of thermal pre-treatment of RB base material

BET specific surface areas (S_{BET}) and pore volumes (V_{pore}) of GACB produced from untreated RB and thermally pre-treated RB at different process temperatures (200^0C to 700^0C), as well as the net Al coating and fluoride removal efficiencies (%) and capacities (mg/g) are presented in Table 5.3.

Chapter 5 Fluoride removal from drinking water using granular aluminum-coated bauxite as adsorbent: Optimization of synthesis process and equilibrium study

177

Thermal pre-treatment of RB prior to Al coating resulted in a strong increase in specific surface areas and pore volumes of GACB, which may be attributed to evolution of water and other sublimable substances in the RB, prior to the Al coating (Kumar et al., 2014). The weight loss upon thermal pre-treatment of RB at different temperatures are given in Table 5.3. The increased surface area may also be a contribution from the conversion of gibbsite in RB to boehmite (upon calcination at temperatures above 200°C), the later which has a higher surface area (Kumar et al., 2014; Das and Das, 2005).

Table 5.3: Surface area, porosity, net Al coating and F removal by RB and GACB produced from untreated and thermally pre-treated RB.

Parameter	RB	GACB (using untreated RB)	GACB (using thermally pre-treated RB)				
			200°C	400°C	500°C	600°C	700°C
Wt loss (%)	-	-	22.8	24.3	28.5	29.2	32
S_{BET} (m²/g)	2.1	0.7	197	181	174	137	120
$V_{pore.}$(cm³/g)	-	0.004	0.122	0.128	0.142	0.151	0.151
Net Al (mg/g)	-	164	219.5	208	392	409.8	307.3
F removal (%)	26.4	67.3	90.6	90	94.1	90	91.6
F removal (mg/g)	0.129	0.335	0.422	0.419	0.426	0.406	0.415

Table 5.3 shows the thermal pre-treatment also markedly improved the net Al coating, hence increased incoproration of gibbsite into GACB. A further increase in fluoride uptake efficiency of GACB from 67.3 % to more than 90 % (and corresponding increased capacity from 0.34 to 0.42 mg F/ g GACB) was also observed. Fig. 5.5 presents the fluoride uptake by GACB produced from thermally pre-treated RB at varying process temperatures.

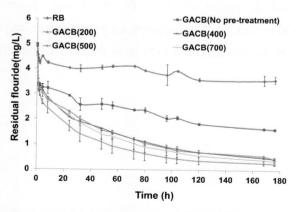

Fig. 5.5 Fluoride uptake by GACB produced from thermally pre-treated RB at different process temperatures (coated at pH 2). Model water: Fluoride = 5 ± 0.2 mg/L, pH = 7.0 ± 0.2, Contact time = 176 h, Temp. = 20°C, GACB dose 10g/L

The further increase in fluoride adsorption may be attributed to the combined effects of increased surface area and pore volume as well as the increased gibbsite content (hence increased surface reactivity/affinity), all of which may influence the adsorption of fluoride ions by GACB (Das and Das, 2005; Cater et al., 1986).

5.3.2.4 Correlations between GACB defluoridation efficiency and surface area, pore volume and Al coating

Pearson's correlation (r), was further used to examine the strength of relationship between GACB defluoridation efficiency (%) and S_{BET} (m²/g), pore volume (V_{pore}(cm³/g)) and Net Al coating (mgAl/g). A strong to very strong positive correlation (Adams et al., 2001) was found, namely; defluoridation efficiency (%) vs. net Al coating (mgAl/g); r = 0.88; defluoridation efficiency (%) vs. S_{BET} ; r = 0.9; and defluoridation efficiency (%) vs V_{pore} ; r = 0.97, at significance levels (p) of < 0.05. The strong correlations further supports the usefulness of the applied working approach for optimizing the synthesis of a fluoride adsorbent, using RB as base material.

Fig. 5.6 compares the fluoride uptakes by RB, GACB coated at pH 2 using thermally untreated RB base material and GACB produced at pH 2 using thermally pre-treated RB at temperature of 500°C, which was considered the optimum synthesis process condition in this study, and therefore selected for further work and analysis. It was however observed from Table 5.3 that, the fluoride uptake efficiency for the various pre-treatment temperatures are similar (i.e. 90 to 94.1 %). Thus in practice a pre-treatment temperature of 200°C could be employed in the production of GACB, as that may be more energy-efficient.

Chapter 5 Fluoride removal from drinking water using granular aluminum-coated bauxite as adsorbent: Optimization of synthesis process and equilibrium study

179

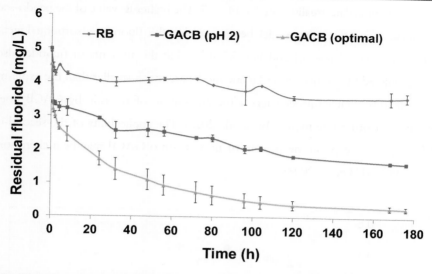

Fig. 5.6: Fluoride removal by RB, GACB produced at coating pH 2 with no thermal pre-treatment and GACB produced at coating pH 2, using thermally pre-treated RB at temperature of 500^0 C (optimal). Model water: fluoride = 5.0 ± 0.3 mg/L, HCO_3 = 260 mg/L, pH = 7 ± 0.2, adsorbent dose = 10 mg/L, room temperature (20°C), Contact time = 176 h, shaker speed = 100 rmp.

5.3.3 Kinetic study of fluoride adsorption onto GACB

To understand the dynamics and underlying mechanisms of fluoride uptake by GACB, the kinetic experimental data were analyzed using four kinetic models, namely, the pseudo-first-order and the pseudo-second-order kinetic models, Bangham and intra-particle diffusion models. Rate constants were calculated using the pseudo-first- and the pseodo-second-order models, the mathematical representations (non-linear form) of which are given are presented in chapter 3, where they were applied for modeling the kinetics of fluoride adsorption onto AOCP. The coefficient of determination (R^2) value was applied to determine which of the kinetic models best fit the experimental data.

The kinetic rate constants, k_1 (h^{-1}) and k_2 (g/(mg h)), the adsorption capacities predicted q_e (cal) and experimental q_e (exp.) at equilibrium, and R^2 values determined based on the pseudo-first-order and pseudo-second-order kinetic models are given in Table 5.4. Comparative plots based on the parameters obtained from the two kinetic models as well as the experimental data are illustrated in Fig. 5.7. The higher R^2 value of the peudo-second-order equation indicates the model better explains the fluoride adsorption kinetics of GACB. Moreover, Table 5.4 and Fig. 5.7 show that the amounts of fluoride adsorbed (q_t(cal)) predicted by the first-order kinetic model are much smaller than the experimental values (q_t(exp.)), which further suggest the adsorption of fluoride by GACB does not follow the first-order-rate model (Bia et al., 2012). The applicability of the second-order-model suggests adsorption mechanism of fluoride onto GACB may be a chemisorption process (Maliyekkal et al., 2008).

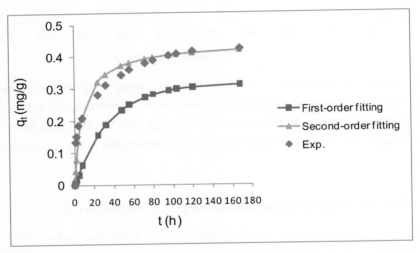

Fig. 5.7 Adsorption kinetics of fluoride onto GACB: Plots for first-order and second-order models, and experimental data.

Even though adsorption onto solid surfaces is generally considered a surface phenomenon, the effect of intra-particle diffusion into porous adsorbents on adsorption processes cannot be ignored, and is necessary to be investigated. Thus similar to the fluoride-AOCP system as presented in chapter 3, the fluoride adsorption kinetic data was fitted to the Weber-and Morris intraparticle diffusion model to explore the contribution of intraparticle diffusion and to determine rate-limiting step for the fluoride-GACB system (Salifu et al., 2013; Maliyekkal et al., 2008; Allen et al., 1989; Weber and Morris, 1963).

*Chapter 5 Fluoride removal from drinking water using granular aluminum-coated
bauxite as adsorbent: Optimization of synthesis process and equilibrium study*

181

Fig. 5.8a shows the intra-particle mass transfer curve for fluoride adsorption by GACB,
which shows also three distinct phases, similar to that of fluoride adsorption onto AOCP.
The first sharper phase (1) where the fluoride uptake was quite rapid may be attributable
to external surface adsorption. This was followed by an intermediate phase (2) that shows
a much slower uptake, which may be due to intraparticle diffusion, until the final plateau
was reached (3) due to adsorption equilibrium.

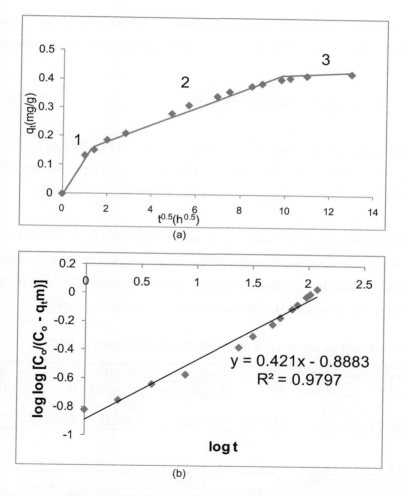

(a)

(b)

Fig. 5.8: (a) Intra-particle mass transfer curve for fluoride adsorption by GACB and (b)
Baghams diffusion model plot.

The k_p value (Table 5.4) was calculated from the slope of a linear plot of q_t against $t^{0.5}$ (not shown). The plot did not pass through the origin, suggesting that even though the adsorption of fluoride onto GACB involved intra-particle diffusion, it was not the only rate-controlling step. The positive value of the intercept, C (Table 5.4) is indicative of some degree of boundary layer control (Salifu et al., 2013; Maliyekkal et al., 2008; Weber and Morris, 1963).

The fluoride-GACB kinetic experimental data was further used to check whether pore diffusion is the only rate-controlling step or not by using Bangham's equation, given as:

$$\log\log\left(\frac{C_o}{C_o - q_t m}\right) = \log\left(\frac{k_o}{2.303V}\right) + \alpha \log(t) \qquad (5.11),$$

where V is the volume of solution (L), m is the weight of GACB used (g/L) and α (< 1) and k_o are constants (Table 5.4). If eqn (5.11) is an adequate representation of the kinetic data, then it indicates the adsorption kinetics is limited by pore diffusion.

The double logarithmic plot according to eqn (5.11) for the fluoride-GACB system (Fig. 5.8b) was found not to satisfactorily fit the experimental data ($R^2 < 0.99$), which supports the earlier finding that the adsorption mechanism involves pore diffusion, but it is not the only rate-controlling step (Haerifer and Azizian, 2013; Mall et al., 2006).

Table 5.4 Kinetic parameters of 4 models for fluoride adsorption by GACB.

First-order-kinetic			Second-order-kinetic			q_e(mg/g) (Exp.)
k_1(1/h)	q_e(mg/g) (cal.)	R^2	k_2(g/mg/h)	q_e(mg/g) (cal.)	R^2	
0.0284	0.316	0.9908	0.2492	0.4420	0.9960	0.4261

Table 5.4 (Continued)

Intra-particle diffusion			Bagham		
k_{ip} (mg/g/h$^{1/2}$)	C(mg/g)	R^2	α	k_o	R^2
0.0289	0.1260	0.9769	0.4210	0.1489	0.9797

Chapter 5 Fluoride removal from drinking water using granular aluminum-coated
bauxite as adsorbent: Optimization of synthesis process and equilibrium study

183

5.3.4 Equilibrium isotherm analysis for GACB

Analysis of equilibrium experimental data is important for determining an isotherm model
that accurately describes the relationship between adsorbed and unabsorbed fractions of
adsorbate, for optimizing the design of adsorption treatment systems. The use of isotherm
models for interpretation of equilibrium data can also provide valuable insights (in
combination with other characterization/interpretation techniques), into the mechanisms
involved in the adsorption process, based on the underlying theories and/or assumptions
in the derivation of the models (Ayoob and Gupta, 2008; Jumasiah et al., 2005; Ho, 2004).
For of the fluoride-GACB system, four frequently used isotherm models; namely, three
two-parameter equations: Langmuir, Freundlich and Temkin models, and one three-
parameter equation: Redlich-Perterson model, were tested to provide the best description
of the adsorption equilibrium data.

The Freundlich model is derived by assuming a heterogeneous surface (multimolecular
layer formation) with a non-uniform distribution of heat of adsorption over the adsorbent
surface. Multimolecular layer formation assumed in the derivation of the Freundlich
isotherm model is associated with physical adsorption process. The nonlinear form of the
Freundilich model is represented by equation (5.12):

$$q_e = K_f C_e^{1/n} \qquad (5.12)$$

where K_f (mg/g) is the Freundlich constant indicative of capacity and $1/n$ is the
heterogeneity factor. The Langmuir and Temkin equations are presented in Table 3.5 of
chapter 3.

The Redlich-Peterson (R-P) isotherm is an empirical model proposed to improve the fit by
the Langmuir or Freundlich equations, and hence combines elements from both models.
The mechanism of adsorption is hybrid, and the model can be applied either in
homogenous or heterogeneous systems. The basic nonlinear form of the R-P model is
expressed as eqn (5.13):

$$q_e = \frac{K_R C_e}{1 + a_R C_e^{\beta}} \qquad (5.13),$$

where K_R (L/g) and a_R (L/mg) are isotherm constants, and $0 < \beta < 1$. The R-P model approaches the Freundlich isotherm at higher adsorbate concentration, and reduces to the Langmuir equation for $\beta = 1$, where $a_R = b_L$ and $q_{max.} = K_R/a_R$. The R-P equation becomes the Henry's equation (eqn (5.14)) when $\beta = 0$:

$$q_e = \frac{K_R C_e}{1 + a_R} \tag{5.14}$$

Equation (5-13) can be linearized as equation (5.15), and by plotting $\ln\left(\frac{K_R C_e}{q_e} - 1\right)$ vs $\ln$ C_e, a straight line will be obtained from which β and a_R can be estimated from the slope and intercept, respectively.

$$\ln\left(\frac{K_R C_e}{q_e} - 1\right) = \beta \ln C_e + \ln a_R \tag{5.15}$$

However, since the R-P equation has three unknown parameters, a nonlinear minimization procedure was adopted by using the Solver add-in function with Microsoft's Excel, and using iterative values of K_R, with the initial value of K_R estimated from the q_{max} and b_L (from the Langmuir isotherm) (Mall et al., 2006; Allen et al., 2003).

Linear regression analysis has frequently been used for determining the best-fit isotherm model, and the method of least squares used for finding the isotherm parameters. The coefficient of determination (R^2) value has often been the only criterion for assessing the quality of isotherm fit to given experimental data. Transformation of the basic nonlinear isotherm equations to the linear forms, however, implicitly alter their error structure, and could also violate the error variance and normality assumptions of standard least squares. This could impact on the appropriateness of using R^2 value as the only criterion in selecting the best-fit isotherm model, and hence the reliability of results. Moreover linear regressions analysis is simple for two-parameter isotherm equations, but not always possible for three-parameter models (Ayoob and Gupta, 2008; Ho, 2004; Allen et al., 2003; Porter et al., 1999). Both linear regression and nonlinear optimization technques were thus employed for determining the best-fit isotherm model and parameters for the fluoride adsorption onto GACB. The isotherms constants and R^2 values determined for the isotherm models described above are presented in Annexes Tables A5.1 to A5.4 and, a summary given on Table 5.5.

Chapter 5 Fluoride removal from drinking water using granular aluminum-coated
bauxite as adsorbent: Optimization of synthesis process and equilibrium study

185

5.3.5 Error analysis and comparison of isotherms

Nonlinear optimization techniques require an error function to be defined to evaluate the fit of the various isotherm equations to the experimental data. The effects of five different error functions were examined for the determination of the best isotherm parameters for the fluoride-GACB system. In each case the parameters were determined by minimizing the respective error functions using the Solver add-in with Microsft's spread sheet, Excel (Microsoft, 2010). The error functions used were: sum of the squares of the errors (ERRSQ), hybrid fractional error function (HYBRID), the average relative error (ARE), Marquart's percentage standard deviation (MPSD) and the sum of absolute errors (EABS), all of which are well presented elsewhere (Kundu & Gunta, 2006; Allen et al., 2003; Porter et al., 1999). The use of different error methods will result in different sets of isotherm constants and error values for each individual isotherm model (Annex 5: Tables A5.1 to A5.4 and Table 5.5), which makes it quite complicated to directly identify the optimum parameter set for each model. The sum of normalized errors (SNE) was therefore used to enable a comparison and identification of the best set of isotherm constants for each model. The SNE values were determined as follows (Kundu & Gunta, 2006; Porter et al., 1999):

- the error values were obtained for each error function for each set of isotherm parameters and were divided by the maximum error value for that error function to obtain the set of normalised errors.

- the normalised errors for each parameter set was summed up to obtain the SNE value.

The obtained SNE values are presented in Annex 5: Tables A5.1 to A5.4 and Table 5.5. The set of isotherm constants for the individual model that gave the minimum SNE value was considered optimum, and was selected as best set of isotherm constants for the model (indicated in bold in Table 5.5). It was observed that, the lowest SNE values for the isotherms were obtained when either the MSPD (Langmuir and Temkin) or ERRQ (Freundlich and R-P) error functions were employed.

Based on R² values as presented in linear transformed (LTFM) column of Table 5.5, three out of the four isothem models appeared to give an acceptable fit to the equilibrium experimental data for fluoride adsorption onto GACB, with a comparative goodness of fit in the order: Freundlich > Temkin > Langmuir. The R-P model, however, exhibited a very poor fit ($R^2 = 0.2701$), which suggested the non applicability of the model to describe the measured equilibrium data.

Nonlinear Chi-square (χ^2) analysis, which has the advantage that all isotherm models can be compared on the same abscissa and ordinate, and also avoids errors associated with linearization, was additionally employed in the determination of the best-fit model. The χ^2 value is given by the mathematical expression as presented equation 4.4 in chapter 4 and is reproduced below:

$$X^2 = \sum_{i=1}^{i=N} \frac{\left(q_{e(exp.)} - q_{e(Cal.)}\right)^2}{q_{e(Cal.)}}$$

The smaller the χ^2 value the more similar the predicted data ($q_{e(cal.)}$) from the isotherm model would be to the experimental data ($q_{e(exp.)}$), hence the more adequate the model can describe the experimental data and vice versa (Ayoob and Gupta, 2008; Ho, 2004). The nonlinear chi - square statistics obtained are shown in Annex 5: Tables A5.1 to A5.4 and Table 5.5.

An inspection of Table 5.5 shows the χ^2 values for all four isotherm model are quite small (0.0202 to 0.0633) and also almost identical. This suggests the predicted data from all the models are quite similar to the experimental data and that all the four isotherm models can reasonably represent the equilibrium adsorption of fluoride onto GACB. This was, however, in contrasts to the results based on the R² values from linear regression which seemed to suggest the non applicability of the of the R-P model. This shows the comparison and selection of best-fit isotherm model based on co efficient of determination (R²) as the sole criterion may not be the most appropriate approach. Therefore based on the nonlinear χ^2 values (Table 5.5), the best-fitting isotherm models for the Fluoride-GACB system was in the order: Freundlich > R-P ≈ Langmuir > Temkin. The applicability

Chapter 5 Fluoride removal from drinking water using granular aluminum-coated bauxite as adsorbent: Optimization of synthesis process and equilibrium study

187

of the Frendlich, Langmuir and R-P models suggest there are both homogenous and heterogeneous binding sites on the surfaces of GACB. These binding sites would be involved in monolayer and multilayer adsorption of the fluoride ions, and the adsorption mechanism may involve both chemisorption and physical adsorption processes (Aziz and Mojin, 2014; Hayes and Mmbaga, 2013; Bansal and Goyal, 2005).

Table: 5.5 Summary of isotherm constants, SNE values, R^2 and x^2 statistics for Langmuir, Freundlich, Temkin and R-P models.

Isotherm Model	Parameters & Statistics	LTFM	ERRQ	HYBRID	MPSD	ARE	EABS
Langmuir	q_{max}(mg/g)	5.6243	14.6289	12.3158	**12.2978**	7.9672	5.6320
	b_L(L/mg)	0.0324	0.0114	0.0136	**0.0137**	0.0210	0.0345
	R^2 (Linear)	**0.9292**	–	–	–	–	–
	x^2	0.0263	0.0284	0.0223	**0.0221**	0.0328	0.0299
	SNE	3.5948	2.5635	2.5576	**2.5071**	4.2960	4.8052
Freundlich	K_f(L/mg)	0.1833	**0.1737**	0.1755	0.1773	0.1563	0.1560
	1/n	0.9023	**0.9357**	0.9282	0.9228	0.9830	0.9847
	R^2 (Linear)	**0.9666**	–	–	–	–	–
	x^2	**0.0202**	0.0207	0.0205	0.0203	0.0297	0.0295
	SNE	3.7575	**3.3879**	3.4177	3.4093	4.4160	4.3581
TEMKIN	A_T(L/mg)	0.7343	0.7343	0.7782	**0.7356**	0.7253	0.7252
	B_T(kJ/mol)	0.6565	0.6565	0.6174	**0.6540**	0.6846	0.6846
	R^2 (Linear)	**0.9447**	–	–	–	–	–
	x^2	0.0588	0.0588	**0.0560**	0.0587	0.0633	0.0633
	SNE	3.7321	3.7314	3.7688	**3.6740**	4.7719	4.7726
R-P	k_R(L/g)	0.2010	**3.7489**	1.5034	0.2144	0.1604	0.1610
	a_R(L/mg)	0.0490	**20.5841**	7.5730	0.3152	0.0314	0.0352
	β	0.9945	**0.067**	0.0799	0.0800	0.3201	0.2989
	R^2 (Linear)	**0.2701**	–	–	–	–	–
	x^2	0.0320	0.0208	**0.0205**	0.0249	0.0295	0.0295
	SNE	4.9203	**3.4186**	3.4435	3.4685	4.9231	4.9201

A comparative fit of the different isotherm (based on the error method that gave the optimum isotherm constants) and the equilibrium experimental data plotted as $q_e = f(C_e)$ is shown in Fig. 5.9.

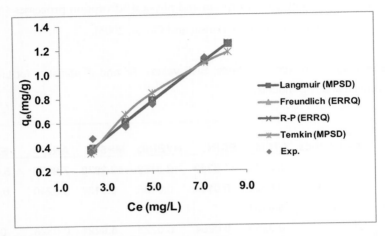

Fig 5.9: A comparative plot for isotherm models and experiment data. Model water: Fluoride = 10 ± 0.2 mg/L, pH = 7.0 ± 0.1, Contact time = 176 h, Temp. = 20°C.

The fluoride adsorption capacity of GACB q_{max} was 12.29 mg/g (Table 5.6) and was also found to be either comparable or fairly higher than the capacities of some reported fluoride adsorbents at pH 6-7. GACB therefore also appear to be promising and could be used for water defluoridation for rural water supply in developing countries especially where bauxite is locally available.

Table 5.6: Comparison of GACB capacity q_{max} (mg/g) with reported fluoride adsorbent capacities.

No.	Adsorbent	q_{max} (mg/g) based on Langmuir model	pH	Reference
1	Granular ferric hydroxide (GFH)	7.0	6-7	Kumar et al., 2009
2	Activated alumina	2.41	7	Ghorai and Pant,2005
3	Sodium exchange montimorillonite-Na$^+$ (ANC-Na$^+$)	1.32		Ramdani et al., 2010
4	Bauxite	5.18	6	Sujana and Anand, 2011
5	Manganese-amended activated alumina	10.12	6.5 -7	Maliyekkai et al., 2008
6	Aluminum (hydr)oxide coated pumice (AOCP)	7.87	7	Salifu et al., 2013
7	Calcined Zn/Al hydrotalcite-like compound (HT1c)	13.43	6	Das et al., 2003
8	**Granular aluminum coated bauxite (GACB)**	**12.29**	**7**	**Present study**
9	Manganese oxide coated alumina	2.85	7	Maliyekkai et al., 2006

5.3.6   Effect of co-existing anions

Natural groundwater may contain other anions that can possibly compete with fluoride for adsorption sites (Kamble et al., 2007). The potential effects of co-anions commonly found in groundwater (nitrate, phosphate bicarbonate, chloride and sulphate) on the efficiency of GACB, were independently examined in batch adsorption experiments using anion concentrations of 2.5mM and GACB dose of 10mg/L (Fig 5.10).

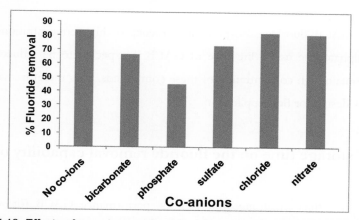

Fig. 5.10: Effects of co-anions on fluoride removal by GACB. Model water: Fluoride = 5 ± 0.2 mg/L, co-ions: 2.5 mM, pH = 7.0 ± 0.1, Contact time = 176 h, Temp. = 20° C, GACB dose 10 g/L

At concentrations of 2.5 mM, phosphate, bicarbonate and sulphate showed negative effect on GACB performance, while nitrate and chloride showed either no or only negligible reduction of fluoride removal efficiency. The decrease of fluoride removal efficiency in the presence of phosphate, sulphate and bicarbonate may be attributable to competition for adsorption sites on the GACB surfaces (Tang et al., 2009; Onyango et al., 2004).

The impact of co-ions on fluoride adsorption by GACB was found to follow the order: phosphate > sulphate ≈ bicarbonate > chloride ≈ nitrate; which may be a reflection of the relative affinities of theses ions for GACB surfaces. The highest impact exhibited by phosphate was presumably due to its adsorption onto GACB surface sites predominantly as an inner-sphere complex, while sulphate and bicarbonate formed both inner-sphere and outer-sphere complexes, and chloride and nitrate forming only outer-sphere complexes with lesser impact (Goldberg, 2010; Tang et al., 2009; Onyango et al., 2004; Su and Suarez , 1997).

Since concentrations of 2.5 mM of phosphate and nitrate are not commonly found in groundwater, additional experiments were conducted with lower concentration of 0.01 mM and16 mM, respectively, which are commonly found in groundwater systems. It was, observed that at phosphate and nitrate concentrations commonly found in groundwater systems, there was only negligible impact on the fluoride removal efficiency of GACB.

Given the reduced GACB fluoride adsorption in presence of high concentrations of bicarbonate and sulphate, the useful life time of GACB is expected to be reduced in groundwater that contain high concentration of these compounds. This therefore has to be considered in any design for field application.

5.3.7 Effect of storage time on the fluoride removal capability of GACB

Similar to AOCP, the effect of storage time on the performance GACB for water defluoridation was assessed in batch adsorption experiments, using freshly produced GACB, and GACB stored for 8 months under normal room conditions (temp. = 20 °C).

Chapter 5 Fluoride removal from drinking water using granular aluminum-coated bauxite as adsorbent: Optimization of synthesis process and equilibrium study

191

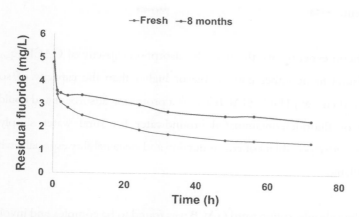

Fig. 5.11 Fluoride removal by GACB stored at different times: adsorbent dose = 10 g/L, particle size = 0.8 - 1.12mm; Model water: fluoride = 5 ± 0.2 mg/L, HCO_3 = 260 mg/L, pH = 7.0 ± 0.2.

Unlike AOCP, however, GACB showed a reduction in its fluoride removal efficiency after 8 months of storage (Fig. 5.11). The factors that contributed to the reduction in the fluorde removal efficiency was not further studied. The preliminary results obtained, however, suggests the storage of GACB for long periods after its production prior to use, may not be recommended.

5.4 Conclusions

Surface modification of bauxite, a locally available material in several countries including Ghana, by aluminium coating was found effective for drinking water defluoridation. Optimal coating conditions comprised pH 2 with thermal pre-treatment of bauxite at 500°C, prior to the coating. Thermal pre-treatment at 200°C is, however, recommended in practice from a production cost point of view, in relation to the small increase of adsorption capacity attained at the higher pre-treatment temperature of 500°C.

Based on sum of normalized errors (SNE) and nonlinear χ^2 analysis, four isotherm models, namely, Freundlich, R-P, Langmuir and Temkin were found capable of describing the equilibrium adsorption of fluoride onto GACB. The adsorption capacity calculated from the models may be helpful in the design of drinking water defluoridation systems based on GACB as an adsorbent.

Based on batch isotherm experiments, the fluoride adsorption capacity of GACB (q_{max} = 12.29 mg/g), was found to be either comparable or higher than the capacity of some reported fluoride adsorbents at pH 6-7. GACB is thus a promising adsorbent and could be used for treatment of fluoride-contaminated groundwater for rural water supply in developing countries, with a possibility of cost reduction and sustainability especially where bauxite is locally available

The mechanism of fluoride adsorption onto GACB was found to be complex and involved both physical adsorption and chemisorption processes, as determined from kinetic and isotherm analyses as well as FTIR and Raman spectroscopic analyses.
Bicarbonate and sulfate were observed to show some retardation on fluoride adsorption capacity of GACB (25 – 35 %), while nitrate, chloride and phosphate showed either no or only negligible effects at concentrations commonly found in groundwater. This has to be considered in any design for field application, for a groundwater source with high concentration of bicarbonate and/or sulfate.

The fluoride removal efficiency of fluoride of GACB was found to reduce after 8 months of storage under normal room conditions.

References

Adams, S., Titus, R., Pietersen, K., Tredoux, G., Harris, C. 2001. Hydrochemical characteristics of aquifers near Sutherland in the Western Karoo, South Africa. J. of hydrology, 241, 91-103.

Alfarra, A.; Frackowiak, E.; Beguin, F., 2004. The HSAB concept as a means to interpret the adsorption of metal ions onto activated carbons. Applied Surf. Sci. 228, 84-92.

Allen, S.J.; Gan, Q.; Mathew, R.; Johnson, P.A., 2003. Comparison of optimized isotherm models for basic dye adsorption by kudzu. Bioresource Technol. 88, 143-52.

Allen, S.J.; Mackay, G.; Khader, K.H, Y. 1989. Intraparticle diffusion of a basic dye during adsorption onto sphagnum peat. Envirron, Pollut. 56, 39-50.

Armenante, P.M. 2009. Adsorption. http://cpe.njit.edu/dlnotes/CHE658/cls11-1.pdf. CitedSeptember 2009.

Ayoob, S.; Gupta, A.K., 2008. Insights into isotherm making in the sorptive removal of fluoride from drinking water. J. Hazard. Mat. 152, 976 -985.

Aziz, H. A.; Mojiri, A., 2014. Wastewater Engineering: Advanced wastewater treatment systems. IJSR Publications, Penang, Malasia.

Bansal, R. C.; Goyal, M., 2005. Activated Carbon Adsorption. CRC Press, Florida.

Baral, S.S.; Das, S.N.; Rath, P.; Chaudhury, G.R., 2007. Chromium (VI) removal by calcined bauxite. Biochem. Eng. J.34, 69 -75.

Bhargava, D.S.; Killedar, D.J., 1991. Fluoride adsorption on fishbone charcoal through a moving media adsorber, Water Res. 26 (6), 781-788.

Bia, G.; Pauli, C.P.; Borgnino, L. 2012. The role of Fe (III) modified montmorillonite on fluoride mobility: Adsorption experiments and competition with phosphate. J. Environ. Management. 100, 1-9.

Black & Veatch Corporation. 2007. Evaluation and assessment of organic chemical removal technologies for New Jersey drinking water. Surface water report. Prepared for New Jersey Advanced Technology, Tarrytown, New York.

Boulard, B., Jacoboni, C., 1989. Raman spectroscopy vibrational analysis of octahedrally coordinated fluoride: Application of Transitions Metal Fluoride Glasses. J. of Solid State Chem., 80, 17-31.

Bratby, J., 1980. Coagulation and flocculation: With an emphasis on water and wastewater treatment. Uplands Press Ltd, Croydon, England.

Cater, D.L.; Mortland, M.M.; Kemper, W.D., 1986. Specific Surface. In Part 1.Physical and Mineralogical Methods-Agronomy Monographs no.9 (2nd Edition); American; American Society of Agronomy--Soil Science Society of America, Madison, pp 413-422.

Chaturvedi, A.K..; Yadava, K.P.; Pathak, K.C.; 1990. Singh, V.N. Defluoridation of water by adsorption on fly ash. Air Soil pollut. 49, 51-61.

Chen, N.; Zhang, Z.; Feng, C.; Suguira, N.; Li, M.; Chen, R., 2010. Fluoride removal from water by granular ceramic adsorption. J. Colloid Interface Sci. 348, 579-584.

Das, D.P.; Das, J.; Parida, K., 2003. Physicochemical characterization and adsorption behavior of calcined Zn/Al hydrotalcite-like compound (HT1c) towards removal of fluoride from aqueous solution. J. of Colloid Inter. Sci. 261, 213-220.

Das, N.; Pattannaik, P.; Das, R., 2005. Defluoridation of drinking water using activated titanium rich bauxite. J. of Colloid and Inter. Sci., 292, 1 - 10.

Dobias, B (Eds)., Coagulation and Flocculation : Theory and applications, Marcel Dekker, Inc. New York, USA. 1993

Fan, X.; Parker, D.J.; Smith, M.D., 2003. Adsorption kinetics of low cost materials. Water Res.37, 4929-4937.

Chapter 5 Fluoride removal from drinking water using granular aluminum-coated bauxite as adsorbent: Optimization of synthesis process and equilibrium study

195

Faust, S.D and Aly, O.M., 1998. Chemistry of water treatment. 2nd ed, CRC Press LLC, Florida, USA.

Fawell, J., Bailey, K., Chilton, J., Dahi, E., Fewtrell, L., Magara, Y. Fluoride in Drinking-water; International Water Association Publishing: London, U.K., 2006.

GES 166/266, Soil Chemistry. Winter, 2000. Lecture Supplement 3. Solid-Water interface; www.GES166_266items_HO-3 Interface.doc.

Ghorai S.; Pant, K.K., 2005. Equilibrium, kinetics and breakthrough studies for adsorption of fluoride on activated alumina. Sep. Purif. Technol. 2,165-173.

Gilbert, B., Stephen, B., Williams, D., G. Mamantov, G., 1988. Raman spectroscopy of fluoride containing chloroaluminate melts. Inorg. Chem., 27(13), 2359 -2363.

Goldberg, S., 2010. Competitive adsorption of molybdenum in the presence of phosphorus or sulfur on gibbsite. Soil Sci., 175(3), 105-110.

Goldberg, S.; Davis, J.A.; Hem, J.D., 1996. The surface chemistry of aluminium oxide and hydroxides, In: Sposito, G (Eds), The Environmental Chemistry of Aluminum. Florida: CRC Press, 271-318.

Gow, N.N.; Lozej, G.P., 1993. Bauxite. Geosciences Canada. 20 (1), 9 - 16.

Grafe, M., Power, G., Klauber, G., 2009. Review of bauxite residue alkalinity and associated chemistry. National Research Flagship. Document DMR-3610, CSIRO: Karawa, Australia.

Groen, J.C.; Peffer, L.A.A.; Perz-Ramirez, J., 2003. Pore size determination in micro-and mesoporous materials: Pitfalls and limitations in gas adsorption data analysis. Mesopor. Mater. 60, 1 - 17.

Haerifar, M.; Azizian, S., 2013. Mixed surface reaction and diffusion-controlled kinectic model for adsorption at the solid/solution interface. J. Phys. Chem. C. 117, 8310 - 8317.

Han, Y.; Park, S.; Lee, C.; Park, J.; Choi, N.; Kim, S., 2009. Phosphate removal from aqueous solution by aluminum (hydr)oxide-coated sand. Korean Society of Env. Eng. 14 (3), 164-169.

Hayes, R. E.; Mabaga, J. P., 2013. Introduction to Chemical Reactor Analysis, Second ed. CRC Press, Florida.

Ho, Y.S., 2004. Selection of optimum sorption isotherm. Carbon. 42, 2115 -2116.

Jodin, M.; Gaboriaud, F.; Humbert, F., 2005. Limitations of potentiometric studies to determine the surface charge of gibbsite γ -Al(OH)3 particles J. of Colloid and Interf. Sci. 287, 581–591.

Jumasiah, A.; Chuah, T.G.; Gimbon, J.; Choong, T.S.Y.; Azni, I. 2005. Adsorption of basic dye onto palm kernel shell activated carbon: sorption equilirium and kinetics studies. Desalination. 186, 57 -64.

Kamble,S.P.; Jagtab,S.; Labhsetwar,N.K.; Thakare, D.; Godfrey,S.; Devotta,S.; Rayalu, A.S., 2007. Defluoridation of drinking water using chitin, chitosan and lanthanum-modofied chitosan. Chem. Eng. J. 129, 173-180.

Kaneko, K., 1994. Determination of pore size and pore size distribution 1. Adsorbents and catalysts, J of Memb. Sci. 96, 59 -89.

Kumar, E.; Bhatnagar, A.; Kumar, U.; Sillanpaa, M., 2011. Defluoridation from aqueos solutions by nano-alumina: Characterization and sorption studies. J. of Hazard. Mat. 186, 1042-1049.

Kumar, E.; Bhatnagar, A.; Ji, M.; Jung, W.; Lee, S.H.; Kim, S.J.; Lee, G.; Song, H.; Choi, J.Y.; Yang, J.S.; Jeon, B.H., 2009. Defluoridation from aqueous solutions granular ferric hydroxide (GFH). Water Res. 43, 490-498.

Kumar, E.; Bhatnagar, A.; Hogland, W.; Marques, M.; Sillanpaa, M., 2014. Interaction of anionic pollutants with Al-based adsorbents in aqueous media - A review. Chem. Eng. J. 241, 443-456.

Kundu, S.; Gupta, A.K., 2006. Arsenic adsorption onto iron oxide-coated cement (IOCC): regression analysis of equilibrium data with several isotherm models and their optimization, Chem. Eng. J. 122, 93–106.

Laveccehia, R.; Medici, F.; Piga, L.; Rinaldi, G.; Zuoro, A., 2012. Fluoride removal from water by adsorption on a high alumina content bauxite. Chem. Eng. Transac. 26, 225-230.

Makov, G., 1995. Chemical Hardness in density functional theory. J. Phys. Chem. 99, 9337-9339.

Maliyekkai, S.M.; Sharma, A. K.; Philip, L., 2006. Manganese-oxide-coated alumina: A promising sorbent for defluoridation. Water Res. 40, 3497 - 3506.

Maliyekkal, S.M.; Shukla, S.; Philip, L.; Nambi, I.M., 2008. Enhanced fluoride removal from drinking water by manganese-amended activated alumina granules. Chem. Eng. J, 140, 183 -192.

Mall, I.D.; Srivastava, V.C.; Agarwal, N.K., 2006. Removal of orange-G and methyl violet dyes by adsorption onto bagasse fly-ash - kinetic study and equilibrium isotherm analysis. Dyes and Pigments. 69, 210 -223.

McBrlde, M.B.; Wessellnk, L.G., 1988. Chemisorption of Catechol on Gibbsite, Boehmite, and Noncrystalline Alumina Surfaces. Environ. Sci. Technol. 22, 703-708.

Mohammad, N. Hydraulics, diffusion and retention characteristic of inorganic chemicals in bentonite. PhD Dessertation, University of South Florida, Flotida, USA. 2004.

Oh, S.J., Cook, D.C., Townsend, H.E., 1998. Characterization of iron oxides commonly formed as corrosion products on steel. Hyperfine Interact. 112(1-4), 59-66.

Okamoto, Y.; Imanaka, T., 1988. Interaction chemistry between molybdena and alumina: Infrared studies of surface hydroxyl groups and adsorbed carbon dioxide on aluminas modified with molybdate, sulfate, or fluoride. J. Phys. Chem. 92, 7102 - 7112.

Onyango, M.S.; Kojima, Y.; Aoyi, O.; Bernardo, E.C., 2004, Adsorption equilibrium model and solution chemistry dependence of fluoride removal from water by trivalent-cation-exchanged zeolite F-9. J. Colloid and Interf. Sci. 279, 341-350.

Pearson, R.G., 2005, Chemical hardness and density functional theory. J.Chem. Sci. 117 (5), 369-377.

Pearson, R.G., 1988. Absolute electronegativity and hardness: Application to inorganic chemistry. Inorg. Chem. 27, 734-740.

Pearson, R.G., 1993. The principle of maximum hardness. Acc.Chem.Res. 26, 25-255.

Porter, J.F.; McKay, G.; Choy, K.H., 1999, The prediction of sorption from a binary mixture of acidic dyes using single- and mixed isotherm variants of the ideal adsorbed solute theory. Chem. Eng. Sci. 54, 5863–5885.

Ramdani, R.; Taleb, S.; Benghalem, A.; Chaffour, N., 2010. Removal of excess fluoride ions from Saharan brackish water by adsorption on natural materials. Desalination. 250,408 - 413.

Rosenqvist, J., 2002. Surface chemistry of Al and Si (hydr) oxide, with emphasis of nano-sized Gibbsite (Al(OH)$_3$. PhD Dissertatin, Umea University, Sweden.

Ruan, H.D., Frost, R.L., Kloprogge, J.T., Schulze, D.G., Duong, L.   FT-Raman spectroscopy and SEM of gibbsite, bayerite, boehmite and diaspore in relation to the characterization of bauxite. Clay Odyssey Conference Proc. 12th International Clay Conference, July 2001, Bahia Blanca, Argentina.

Sahli, M.A.M.; Annouar, S.; Tahaikt, M.; Mountadar, M.; Soufiane, A.; Elmidaoui, A., 2007. Fluoride removal for underground brackish water by adsorption on the natural chitosan and by electrodialysis. Desalination. 212, 37-45.

Salifu, A.; Petrusevski, B.; Ghebremichael, K..; Buamah, R.; Amy, G.L., 2012. Multivariate statistical analysis for fluoride occurrence in groundwater in the Northern Region of Ghana. J. of Contaminant Hydro. 140-141, 34-44.

Salifu, A.; Petrusevski, B.; Ghebremichael, K..; Modestus, L.; Buamah, R.; Aubry, C.; Amy, G.L., 2013. Aluminum (hydr)oxide coated pumice for fluoride removal from drinking water: Synthesis, quilibrium, kinetics and mechanism. Chem. Eng. J. 228, 63 -74.

Sarpola, A., 2007. The hydrolysis of aluminum, a mass spectrometric study, PhD Thesis, University of Oulu, Oulu, Finland.

Sawyer, C.N., McCarty, P.L., Parkin, G.F. 1994.   Chemistry for Environmental Engineering. 4th ed. McGraw-Hill, Inc, New York, 93-98 pp.

Su, C., Suarez, D.L., 1997. In situ infrared speciation of adsorbed carbonate on aluminum and iron oxides. Clays and Minerals, 45(6), 814 -827.

Sujana, M.G.; Anand, S., 2011. Fluoride removal studies from contaminated groundwater by using bauxite. Desalination.267, 222-227.

Tang,Y.; Guan, X.; Wang, J.; Gao, N.; McPhail, M.R.; Chusuei, C.C., 2009. Fluoride adsorption onto granular ferric hydroxide: Effects of ionic strength, pH, surface loading, and major co-existing anions. J. Hazard. Mat.171, 774-779.

Thoma, P.V., Ramakrishman, V., Vaidvan, V.K., 1989. Oxidation studies of aluminum thin films by Raman Spectroscopy. Thin films, 170 (1), 35 -40.

U.S Army Corps of Engineers. 2001. Engineering and Design: Adsorption design guide. http://www.bdg.org/ccb/ARMYCOE/COEDG/dg_111012.pdf.

Vigneresse, J.L.; Duley, S.; Chattaraj, P.K., 2011. Describing the chemical character of a magma. Chem. Geol. 287, 192-113.

Weber, W.J.; Morris, J.C., 1963. Intraparticle diffusion during the sorption surfactants onto activated carbon. J. SanitaryEng. Div., AESC. 89, 53-61.

Wiese, G.R.; Healy, T.W., 1974. Adsorption of Al(III) at the TiO2 -H2O interface. J. of Colloid and Interf. Sci. 51 (3), 434 - 442.

Yan-Pang, W.; Ren-Kou, X.; Jiu-Yu, L, 2013.Effect of FE/Al Hyroxide on desorption of K^+ and NH_4^+ from two soils and Kaolinite, Pedospere. 23 (1), 81 -87.

Zafar, M.N; Nadeem, R; Hanif, M.A., 2007. Biosorption of nickel from protonated rice bran, J. Hazard. Mater. 143 478–485.

Chapter 5 Fluoride removal from drinking water using granular aluminum-coated
bauxite as adsorbent: Optimization of synthesis process and equilibrium study

201

Annex 5

Isotherm constants and error values for individual isotherm models

Table A5.1: Langmuir Isotherm constants, SNE values, R^2 and χ^2 statistics.

Parameters	LTFM (ERRQ)	ERRQ	HYBRID	MPSD	ARE	EABS
q_{max}	5.6243	14.6289	12.3158	12.2978	7.9672	5.6317
b_L	0.0324	0.0113	0.0136	0.0137	0.0210	0.0345
R^2 (Linear)	0.9292					
χ^2	0.0263	0.0284	0.0222	0.0221	0.0328	0.0299
ERRSQ	0.0172	0.0102	0.0103	0.0103	0.0206	0.0178
HYBRID	0.6277	0.4964	0.4942	0.4967	0.7063	0.8334
MPSD	1.2810	1.7904	1.4236	8.28E-05	12.0804	15.8205
ARE	7.8402	5.9940	6.0858	6.2487	6.5786	8.4719
EABS	0.3073	0.1999	0.2015	0.2068	0.2792	0.2888
ERRSQ	0.8355	0.4969	0.5006	0.5008	1	0.8654
HYBRID	0.7529	0.5955	0.5929	0.5958	0.8473	1
MPSD	0.0810	0.1132	0.0910	5.233E-06	0.7636	1
ARE	0.9254	0.7075	0.7184	0.7376	0.7765	1
EABS	1	0.6505	0.6558	0.6729	0.9086	0.9400
SNE	3.5947	2.5635	2.5576	2.5071	4.2960	4.8052

Table A5.2: Freundlich Isotherm constants, SNE values, R^2 and χ^2 statistics.

Parameters	LTFM (ERRQ)	ERRQ	HYBRID	MPSD	ARE	EABS
K_f	0.1833	0.1737	0.1755	0.1773	0.1563	0.1560
$1/n$	0.9023	0.9357	0.9228	0.0923	0.9830	0.9847
R^2 (Linear)	0.9666					
χ^2	0.0202	0.0207	0.0205	0.0203	0.2967	0.0296
ERRSQ	0.0105	0.0097	0.0097	0.0098	0.0123	0.0121
HYBRID	0.4796	0.4666	0.4646	0.4654	0.5924	0.5898
MPSD	1.0369	0.3701	0.6223	4.47E-05	11.1063	10.8348
ARE	6.7183	6.1059	6.1221	6.2516	4.7430	4.7154
EABS	0.2314	0.2018	0.2020	0.2072	0.1643	0.1612
ERRSQ	0.8545	0.7861	0.7929	0.7975	1	0.9884
HYBRID	0.8096	0.7876	0.7843	0.7856	1	0.9956
MPSD	0.0934	0.0333	0.0560	4.026E-06	1	0.9756
ARE	1	0.9089	0.9113	0.9305	0.7050	0.7019
EABS	1	0.8721	0.8732	0.8956	0.7100	0.6966
SNE	3.7575	3.3879	3.4177	3.4093	4.4160	4.3581

Table A5.3: Temkin Isotherm constants, SNE values, R^2 and χ^2 statistics.

Parameters	LTFM (ERRQ)	ERRQ	HYBRID	MPSD	ARE	EABS
B_T	0.6565	0.6565	0.6174	0.6540	0.6846	0.6846
A_T	0.7343	0.7343	0.7782	0.7356	0.7253	0.7252
R^2 (Linear)	0.9447					
χ^2	0.0588	0.0588	0.0560	0.0587	0.0632	0.0633
ERRSQ	0.0364	0.0364	0.0391	0.0364	0.0410	0.0410
HYBRID	1.4551	1.5449	1.3785	1.4453	1.6932	1.6935
MPSD	0.5797	0.5738	2.00E-07	1.15E-06	9.3326	9.3361
ARE	10.5045	10.5044	10.5437	10.5231	10.0230	10.0231
EABS	0.4123	0.4124	0.4453	0.4156	0.3661	0.3660
ERRSQ	0.8886	0.8886	0.9548	0.8893	0.9999	1
HYBRID	0.8592	0.8591	0.8140	0.8534	0.9999	1
MPSD	0.0621	0.0615	2.1459E-08	1.2282E-07	0.9996	1
ARE	0.9963	0.9963	1	0.9980	0.9506	0.9506
EABS	0.9259	0.9260	1	0.9333	0.8220	0.8220
SNE	3.7321	3.7314	3.7688	3.6740	4.7720	4.773

Table A5.4: R-P Isotherm constants, SNE values, R^2 and χ^2 statistics.

Parameters	LTFM (ERRQ)	ERRQ	HYBRID	MPSD	ARE	EABS
K_R	0.2010	3.7489	1.5034	0.2144	0.1604	0.1610
a_R	0.0490	20.5841	7.5730	0.3152	0.0314	0.0352
β	0.9945	0.0670	0.0800	0.0800	0.3202	0.2989
R^2 (Linear)	0.2701					
χ^2	0.0320	0.0208	0.0205	0.0249	0.0295	0.0295
ERRSQ	0.0215	0.0097	0.0097	0.0119	0.0121	0.0121
HYBRID	1.1503	0.6222	0.6200	0.7214	0.7856	0.7849
MPSD	11.3710	0.4362	0.7014	6.91E-07	12.3580	12.3391
ARE	9.3929	6.1052	6.1190	6.1413	4.7367	4.7384
EABS	0.3519	0.2018	0.2018	0.2152	0.1619	0.1620
ERRSQ	1	0.4482	0.4521	0.5518	0.5623	0.5618
HYBRID	1	0.5410	0.5389	0.6272	0.6830	0.6824
MPSD	0.9201	0.0353	0.0568	5.595E-06	1	0.9985
ARE	1	0.6500	0.6514	0.6538	0.5043	0.5045
EABS	1	1.7441	1.7443	1.6357	2.1735	2.1730
SNE	4.9201	3.4186	3.4434	3.4685	4.9231	4.9201

6

Aluminol (Al-OH) fuctionalized wood charcoal for treatment of fluoride-contaminated groundwater: Effect of wood source, particle size, surface acidity-basicity and field assessment

Main part of this chapter is under preparation for publication in Carbon Journal.

Abstract

Aluminum oxide coated media (AOCM) have been found capable of water defluoridation. The fluoride removal performance of AOCM, however, require further enhancement in order to improve on the economic and practical viability, for sustainability. AOCM defluoridation materials were produced by surface modification of indigenous materials by aluminium coating and, so far using bauxite and pumice as base materials. The specific surface areas of AOCMs produced from the pumice and bauxite base materials were, however, found to be low which may be attributable to the low specific areas of these precursor base materials. In this component of the study, the possibility of using wood charcoal (WC) as an alternate indigenous base material for producing a high-surface-area fluoride adsorbent, with enhanced performance was investigated. Aluminol functionalized wood charcoal (AFWC) were found capable of reducing fluoride concentration of 5 ± 0.2 mg/L in model to ≤ 1.5 mg/L, in laboratory-scale columns. The number of bed volumes of water treated by AFWC was found to be 30% higher than that of aluminium oxide coated pumice (AOCP). The superior performance of AFWC over that of AOCP may be due to differences in the textural properties (surface area and porosity) of the precursor base materials (virgin pumice versus wood charcoal), used in the fluoride adsorbent synthesis process. It was further observed that, different starting/precursor wood charcoal (i.e. WC 1, WC 2, and WC 3 & WC 4) used as base marerial for the surface functionalization process had an influence on the fluoride adsorption capabilities of the produced AFWCs. Four aluminol functionalized wood charcoals (AFWCs) were produced: AFWC 1, AFWC 2, AFWC 3 and AFWC 4, using WC 1, WC 2, WC 3 and WC 4 as precursor wood charcoal, respectively. Their fluoride removal efficiencies as well as kinetics in terms of contact times for reducing the fluoride concentration in model water to the WHO guideline value was in the order: AFWC 1< AFWC 2 < AFWC 3 < AFWC 4. AFWC 3, which was used for further work, was found capable of reducing fluoride concentration of 4.88 mg/L in natural groundwater in the field to ≤ 1.5 mg/L (WHO guideline value), indicating its laboratory performance was reproducible under field conditions. A comparison of the performance of AFWC with that of activated alumina (AA) under similar field conditions, indicated a higher performance of the later. The fluoride adsorption capacity of AFWC therefore require further improvements. The particle size range of the AA grade used for the field assessment was, however, much finer (0.21 – 0.63 mm) than that of AFWC (0.8 – 1.12mm), which might have contributed to its better performance. Moreover the grade of AA tested was said not be regenerable and has to be used only once and disposed off. The use AA for treatement of fluoride-contaminated groundwater in developing countries

Chapter 6 Aluminol (Al-OH) fuctionalized wood charcoal for treatment of
fluoride-contaminated groundwater: Effect of wood source and field assessment

205

can, however, only be cost-effective when it can be used for multiple cycles. On the other hand AFWC was found to be regenerable. Regenerated AFWC (RAFWC) was further observed to perform by 30% better in the field than the freshly produced AFWC. The field performance of AFWC and RAFWC were therefore found encouraging and it could be too early to conclude on the superiority of the performance of AA over that of AFWC, since the former (i.e. AA) can not be regenerated.

6.1   Background

Study results as presented in Chapters 3 to 5, have revealed Aluminium Oxide Coated Media (AOCM), as a promising defluoridation material, which could be used for water treatment in developing countries. The fluoride removal performance of AOCM with regards to the number of bed volumes (BV) of water treated however require further enhancement in order to improve on its economic and practical viability, hence sustainability when in use. AOCM defluoridation materials were produced by surface modification of indigenous materials through an aluminium coating process, and so far using bauxite and pumice as base material. The specific surface area of AOCM produced from pumice and bauxite base materials were, however, found to be low (AOCP: $S_{BET} = 1.5 \ m^2/g$ and for GACB: $S_{BET} = 0.7 \ m^2/g$), which may be attributable to the low specific areas of their respective precursor base materials, that is for virgin pumice, $S_{BET} = 3.4 \ m^2/g$, and raw bauxite (RB), $S_{BET} = 2.1 \ m^2/g$. An attempt to thermally pre-treat virgin pumice prior to the aluminium coating process in order to improve on the surface area of produced AOCP was not successful due to the reduced mechanical strength after the calcination process and the procedure was therefore discontinued. As presented in section 3.3.3 (chapter 3), direct thermal treatment of AOCP (i.e after the Al coating) also aimed at improving the defluoridation performance instead affected the performance negatively. Presumably due to its good mechanical strength, however, thermal pre-treatment of raw bauxite (RB) prior to the Al coating process was successful in improving both the specific surface area and the surface reactivity of the coated material, and the combined effect resulted in a material (GACB) with higher fluoride adsorption capability, as presented in section 5.4.2.3 of chapter 5. The results therefore suggested that the specific surface area of a base material may be one of the important properties to be considered in its use for synthesizing aluminium oxide coated media (AOCM) for water defluoridation. In practice, however, thermal pre-treatment of the base material would have to be carried out under controlled conditions, which may require the use of specialised equipment such as a muffle furnace and would also result in energy cost, all of which may increase the production cost of the defuoridation material. In that context the use of an indigenous base material with high surface area which could eliminate such potential production cost components becomes of interest. Wood charcoal is one such indigenous material that can possibly be

Chapter 6 Aluminol (Al-OH) fuctionalized wood charcoal for treatment of
fluoride-contaminated groundwater: Effect of wood source and field assessment

207

used to produce a fluoride adsorbent with high-surface-area. Wood charcoal is rather cheaper, more readily available and easily accessible in most developing countries compared to bauxite and pumice. More over wood charcoal can be renewable material through tree planting/afforestation and can therefore be used sustainably (Nishimiya et al., 1998), compared to pumice and bauxite which are non-renewable resources.

Wood charcoal (WC) is known to have adsorption properties which has been used during millennia for different applications. From a historical perspective, the use of wood charcoal for purification processes dates back to ancient times. For instance in Egyptian papyri dating from 1550 BC, the use of charcoal for adsorption of odorous vapor from putrefying wounds and intestinal tracks was mentioned. The ancient Greek also used charcoal for treating food poisoning through adsorption of the toxins emitted by ingested bacteria, thereby reducing the toxic effects. Hindu documents dating back from 450 BC, particularly mention the use of charcoal filters for drinking water purification by the adsorption process (Cecen and Aktas, 2011; Pastor-Villegas et al., 2006).

WC is obtained from raw wood by carbonization, a process which converts the raw wood (RW) through pyrolysis in an inert atmosphere (absence of oxygen) into a black solid carbonaceous material, composed mainly of carbon atoms, heteroatoms (mainly oxygen, nitrogen) and mineral matter as ash. The starting/precursor raw wood material is known to be one of the factors that has an influence on the characteristics of the produced wood charcoal (Pastor-Villegas et al., 2006; Czernik, 2008). The surface of wood charcoal consist of graphene sheets, which are in the form of one-atom-thick layers of pure carbon, and on the edges of the graphene sheets are the heteroatoms which form inherent functional groups on the charcoal particle surfaces. These include oxygenated groups such as carbonyl, carboxylic and phenolic hydroxyl groups as well as nitrogen groups such as pyridone, pyrrole and pyridine, (as present schematically in Fig. 6.1, and are responsible for the characteristic surface reactivity of wood charcoal particles (Pastor-Villegas et al., 2006; Thakur & Thakur, 2016; Beguin & Frackowiack, 2010; Figueiredo, 2013). The adsorption of aqueous species by carbonaceous materials such as wood charcoal is predominantly due to the formation of surface complexes between the target species and the functional groups on the particle surfaces.

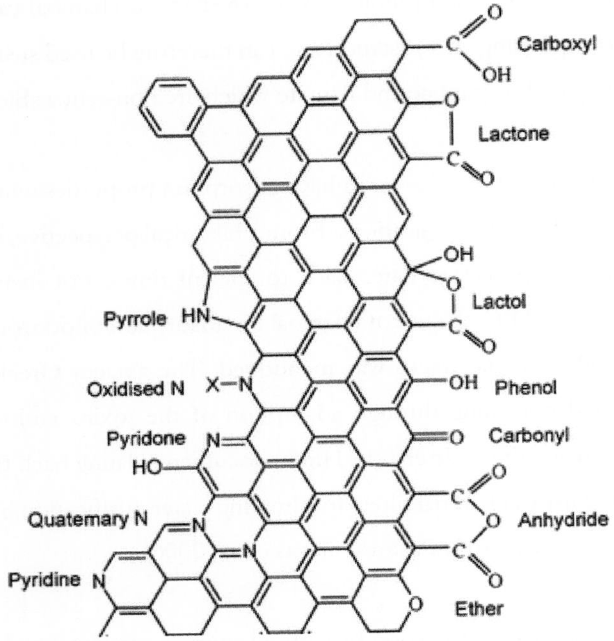

Fig. 6.1 Schematic diagram of functional groups (nitrogen and oxygen surface groups) on carbon. (Source: Figueirido (2013)).

Even though carbonaceous materials are known to possess adsorption capabilities for a range of contaminants due to their characteristics such as, high specific surface area, pore size distribution, pore volume and presence of surface functional groups, they are also observed to be less effective for the removal of certain contaminants, especially inorganic species present in aqueous solutions. For instances, attempts to apply activated carbon or different charcoal materials for fluoride removal from drinking water in particular, showed very low defluoridation potentials (Gupta and Ayoob, 2016; Yin et al al., 2007; Bhatnager et al., 2011). The mechanism of fluoride removal by activated carbon was observed to be governed primarily by a physical adsorption process, with the surface area playing an important role. It was further found that the inherent functional groups such as phenolic hydroxyl groups and carboxyl groups on the carbonaceous material surfaces, played no role in the removal of fluoride ions from aqueous solution. The low fluoride adsorption

Chapter 6 Aluminol (Al-OH) fuctionalized wood charcoal for treatment of
fluoride-contaminated groundwater: Effect of wood source and field assessment

209

tendency was ascribed to the nonmetallic nature of carbonaceous materials, as compared to metallic materials such as activated alumina which are known to process good affinity for aqueous fluoride ions (Gupta and Ayoob, 2016).

Based on the chemistry of a given adsorbate, however, the nature of the surface groups on a given adsorbent can be tailored by its modification/functionalization, for adsorption of the target adsorbate (Ma et al., 2012; Figueiredo, 2013; Daley et la., 1996; Ahmed et al., 2011). Surface treatments of wood charcoal (WC), in other to modify the nature of the inherent surface functional groups to have good affinity/reactivity specifically for fluoride, is therefore of interest in the search for appropriate defluoridation materials. In this context, an aluminol (AlOH)-based surface functionalization approach is still considered to be a more appropriate method for enhancing the fluoride adsorption potential of WC, in accordance to the hard soft acid base (HSAB) concept, as discussed in chapter 3 (section 3.1) of this thesis. Such modification of the surface chemistry of WC would also take advantage of its good specific surface area, the combination of which could result in the synthesis of a novel fluoride adsorbent with a good potential for field application.

One of the factors that influences the adsorption properties of a given adsorbent material is the particle size range. The rate of adsorption is known to be inversely related to the particle size (Muller, 2010). In a study conducted by Canales et al. (2013), for adsorption of arsenic using laterite, it was found that the particle size influenced both the kinetics and equilibrium characteristic of adsorption of the species. The equilibrium adsorption capacity was found to increase from 100 mg As/Kg to over over 200 mg As/Kg (more than twice), when the laterite particles coarser than 4 mm was reduced to particles finer than 75 µm. This indicated the extent of influence of particle size of materials on their adsorption properties, eventhough too fine particles may not be suitable for application in column filters due to potential problems associated with clogging.

The adsorption process involve interactions of the adsorbent surfaces with its environment, which are mainly governed by acid-base interactions. The acidity-basicity properties of an adsorbent material is therefore a major determinant which defines its adsorption behavior. An understanding or studies related to the acidic and basic nature of an adsorbent surface thus remain of research interest in the development of defluoridation materials (Gupta and Ayoob, 2016; Alemdaroglu, 2001).

According to Bhatnager et al. (2011), some synthesized adsorbents may show good fluoride removal capability in laboratory batch conditions but, however, fail under field conditions, which makes the search and/or selection of an appropriate fluoride adsorbent a tedious process. This therefore underscores the critical need for assessing the efficacy of a synthesized fluoride adsorbent, which may show promising results in the laboratory, also under field conditions, in order to determine its potential for treating natural fluoride-contaminated groundwater/practical application. As also discussed in chapter 4 (section 4.1), the regenerability of a given fluoride adsorbent remain an important factor that may contribute to its economic viability and practical application.

The aims of this study were therefore to: (i) explore the possibility of providing aluminol (AlOH) functionality to wood charcoal by incorporation of metallic species (Al (III)) into its fabrics in order to tailor it for fluoride adsorption; (ii) investigate the effect of the precursor wood charcoal (WC) on the fluoride adsorption potential of the Aluminol functionalized wood charcoal (AFWC), (iii) assess the performance of AFWC under continuous flow conditions, (iv) investigate the regenerability of exhausted AFWC; (v) study the acidity-basicity of AFWC and regenerated AFWC (RAFWC) surfaces in relation to their fluoride adsorption behaviour; (vi) assess the efficacy of both AFWC and RAFWC for water defluoridation in the field in pilot-scale columns, using natural fluoride-contaminated groundwater, and (vii) conduct for comparison similar field pilot-scale column study with the industry standard, activated alumina (AA).

6.2   Materials and Methods

6.2.1 Synthesis of Aluminol (AlOH) functionalised wood charcoal (AFWC)

6.2.1.1 Production of wood charcoal

Samples of wood charcoal used for the study were produced locally in Ghana using the traditional earth mound technology. The wood to be carbonized was cut to size and stacked

Chapter 6 Aluminol (Al-OH) fuctionalized wood charcoal for treatment of
fluoride-contaminated groundwater: Effect of wood source and field assessment

211

together on the ground in the form of a mound or pile. This was well covered with earth material that formed an air-tight insulating barrier, thus providing inert conditions for the carbonization process. The inert condition was necessary to prevent the charcoal from burning into ash. An initial combustion of a small portion of the wood pile, provided enough heat required for carbonization of the rest. Four different sample of charcoal were produced, WC 1, WC 2, WC 3 and WC 4, using wood from four selected trees.

6.2.1.2 Preparation of Aluminol fictionalized wood charcoal (AFWC)

Bulk wood charcoal samples were transferred to the UNESCO-IHE laboratory, where they were crushed and sieved to two different particle size ranges; small size range: 0.425 – 0.8 mm and large size range: 0.8 – 1.12 mm. These were washed, air-dried and used for the surface modification process. Surface treatment in order to provide aluminol (AlOH) functionality to the wood charcoal particles was accomplished an aluminium coating/incorporation and neutralization process, using 0.5 M $Al_2(SO_4)_3$ and 3 M NH_4OH respectively, similar to the procedure described in previous chapters (3 & 4). Four aluminol functionalized wood charcoals were prepared: AFWC 1, AFWC 2, AFWC 3 and AFWC 4 and used for subsequent fluoride adsorption studies. Activated alumina used for field pilot study was provided by a supplier from Canada.

6.2.1.3 Physico-chemical of characterization wood charcoal and AFWC

Fourier transform infrared (FTIR), scanning electron microscopy (SEM), EDX spot elemental analysis, x-ray fluorescence (XRF), BET specific surface area characterization as well as pore size distribution analysis were conducted using similar procedures described in chapters 3 (section 3.3.1) and 5 (section 5.3.2).

6.2.1.4 Aluminum stability/leaching experiment

Due to the influence of pH on Al mobility (Reitzel et al., 2013), an experiment was conducted to check any possible leaching of the aluminium incorporated in AFWC, under varying pH conditions. This was to assess the stability of the Al incorporated into AFWC, as any leaching could deteriorate the quality of the produced defluoridated water. Series of

100 ml PE bottles were each filled with 50 ml of demineralized water that was buffered with sodium bicarbonate. The pHs of the buffered water in each of the PE bottles were adjusted with NaOH and/or HCl to desired values between 5 and 9 (i.e. 5.0, 5.5, 6.0, 6.5, 7.0, 7.5, 8.0, 8.5 and 9.0) for the leaching test. One gram (1 g) of AFWC was added to the contents of each PE bottle, and the bottles were firmly closed to avoid the influence of atmospheric CO_2, during the leaching experiment. The PE bottles were placed on a shaker for 24 h, after which samples of the solution were taken and analysed for amount of aluminium leached from AFWC. The leaching experiment for each pH value was done in duplicate.

6.2.1.5 Fluoride adsorption experiments using wood charcoal and AFWC

The fluoride adsorption potentials of the four produced AFWCs were initially assessed and compared through batch kinetic adsorption mode experiments, using similar procedures described in previous chapters (3 and 5). The most promising among them was selected for further assessment in similar laboratory-scale column experiments, as described in chapter 4 (section 4.2.3). A control fluoride adsorption experiment was also conducted using the virgin wood charcoal base material.

6.2.1.6 Regeneration of exhausted AFWC

Regeneration was accomplished by re-incorporating Al^{3+} into the exhausted AFWC, thereby providing a new aluminol (AlOH) functionality as new active sites for restoring the fluoride adsorption capacity. The conceptual approach and procedure was similar to that applied for the regeneration of AOCP (section 4.2.4). Thus after running the column experiment with AFWC till exhaustion, the fluoride-saturated AFWC was re-soaked in sufficient amount of 0.5 M $Al_2(SO_4)_3$ solution in order to re-incorporate Al^{3+} for providing the aluminol functionality. The AFWC (RAFWC) was subsequently used for fluoride adsorption studies under similar conditions as that of freshly prepared AFWC.

Chapter 6 Aluminol (Al-OH) fuctionalized wood charcoal for treatment of
fluoride-contaminated groundwater: Effect of wood source and field assessment

213

6.2.1.7 Potentiometric titration

The acidity-bacisity of the AFWC and RFWC surfaces were determined by potentiometric
titration techniques. The titration was perforemed by using Titrano Plus (Metrohm,
Switzerland). Samples of each adsorbent were measured (1.5 g), and added to 0.01M
$NaNO_3$ (as electrolyte solution) in a container, and allowed to equilibrate over 24 h, after
which the titration was conducted. In order to eliminate the influence of CO_2, the
suspension in the container was continuously saturated with N_2, througt the titration. 0.1
M NaOH was used for the titrant, and the titration was perfomed starting from the initial
pH of the suspension. Measurement were done for the pH range of 5.5 – 10, which
encompassed the pH at which the fluoride adsoption potential of AFWC and RFWC were
assessed, i.e pH 7.

6.2.1.8 Field pilot-scale study in Ghana

Field assessment of the efficacy of the produced adsorbent for defluoridation of natural
fluoride-contaminated groundwater was conducted in the Bongo town in the Upper East
region of Ghana, which is known to be within one of the most fluoritic areas in the country.
The study focused on operation of household filters that produces water for drinking and
cooking only, the fraction of the total water demand that can cause incidence of fluorosis
and other related health hazards.

Simple "point of use" household defluoridation units (HDUs) were employed for the pilot
study. The defluoridation units comprised of five main components: feed/raw water and
treated water tanks, PVC column filled with fluoride adsorbent, an orifice, a frame/housing
and a system of tubings/valves. A PVC pipe, 10 cm in diameter and 62 cm long was housed
in a locally fabricated metal frame support. A 25 L locally purchased plastic bucket used as
the raw water tank, was placed on the metal support, and arranged above the PVC column,
with adequate hydraulic head as to allow operation of the system under gravity. A second
25 L plastic bucket fitted with an outlet tap was placed on on a lower stand to collect the
treated water (Fig 6.2). The raw water tank, PVC column and treated water tank were
connected through a system of tubings/valves. Both raw water and treated water tanks
were well covered to avoid contamination and the top covers were perforated with small

holes for ventilation. The PVC pipe, was packed with the adsorbent to be tested to a filter bed of height 37 cm. An 8 cm depth of graded gravel/coarse sand support was provided at the bottom of the column, and 5 cm of supernatant water was allowed at the top. The feed water was allowed to flow through the column in an up-flow mode, with an orifice installed in the PVC top cap arrangement for controlling/maintaining the flow as designed. The tubing and valve arrangement were designed to allow for draining and backwashing, when required. The pilot household defluoridation units were designed to a treat a typical Northern Ghana 10-member household/family water demand for drinking and cooking of 50 L per day. The units were designed to operate for 20 h per day (that is a design flow of 41.7 ml/min). Three household defluoridation units were employed: two for testing the efficacies of AFWC and RAFWC, and the third for testing fluoride removal capacity of activated alumina (AA) for comparison (Fig 6.2).

The units were installed in a selected family home close to a fluoride-contaminated groundwater source (a borehole which is also source of drinking water for the family, even though with high fluoride concentration). The feed water tank were filled on average twice a day for its operation (mostly in the morning and evening). Samples of the filtrate from the units were collected at periodic intervals and analysed for residual fluoride.

Chapter 6 Aluminol (Al-OH) fuctionalized wood charcoal for treatment of fluoride-contaminated groundwater: Effect of wood source and field assessment

215

Fig. 6.2 Pilot-scale household defluoridation units at Bongo, Upper East region of Ghana, filled with AFWC, RAFWC and AA.

6.3 Results and Discussion

6.3.1 Characterization of wood charcoal (WC) and AFWC

The elemental composition of both wood charcoal (WC) and aluminol functionalized wood charcoal (AFWC), obtained from XRF and EDX analyses are presented in Table 6.1.

Table 6.1 Elemental composition (by wt %) of WC and AFWC obtained from XRF and EDX.

Element		C	O	Na	Al	Si	K	Fe	S	Ca
EDS (wt %)	WC	91.8	7.1	-	-	-	0.3	-	-	0.3
	AFWC	66	13.9	-	11.7	-	-	-	4.8	2.2
XRF (wt %)	WC	-	-	1.7	1	1.7	10.9	0.65	1.1	51.7
	AFWC	-	-	1	58.3	0.53	1	0.96	17.3	12.4

The XRF results showed that aluminium (Al) was present in WC only in trace amounts (1%). After the surface treatment, however, the presence of Al increased to 58 %, indicating the incorporation of the metal into the wood charcoal material. The EDX spot analysis gives the elemental composition of samples to a depth of several micrometers. Based on the average value for two spot analysis, the presence of carbon (C), in wood charcoal was 91.8 %, which was, as expected, the dominant element, followed by oxygen (7.1 %), while Ca an K were in trace amounts (Table 6.1). After the surface treatment/functionalization process, the EDX analyses (average of 3 spots) of the produced material (AFWC) showed a decrease in the presence of carbon (66 %), while oxygen increased to 13.9 %, with an emergence of Al (11.7 %). The EDX and XRF analyses thus supported each other with regards to the incorporation of Al into the wood charcoal base material.

The SEM images of the wood charcoal (WC 3) before (a) and after (b) the surface treatment (AFWC 3) are presented in Fig. 6.3a and Fig. 6.3 b, respectively. The SEM image of wood charcoal (WC 3) particles (Fig 6.3 a) appeared like an accumulation of tubes joined to each other. This type of structure gives the particle a high porosity, hence a high surface contact area, suggesting the suitability of wood charcoal as a base material for the incorpation of Al in the surface functionalization process and, the subsequent application for fluoride adsorption.

After the surface treatment process using the aluminium - bearing solution (Fig. 6.3 b), the sample exhibited particles attached to the surfaces and also inside inside the grain pores (see Fig. 6.3b, x 1,000), which are presumably aluminium oxides, thus supporting the observation from the XRF and EDS analysis. The presence of Al in the form of the oxides after the surface treatment process, is expected to provide the required aluminol (AlOH) functionality to wood charcoal for enhancing its fluoride removal capability. The process of formation of the surface hydroxyl (i.e AlOH) functional groups is presented in section 3.3.7 of the thesis (including related references), with schematic illustrations using Figs 3.10 & 3.11.

Chapter 6 Aluminol (Al-OH) fuctionalized wood charcoal for treatment of fluoride-contaminated groundwater: Effect of wood source and field assessment

217

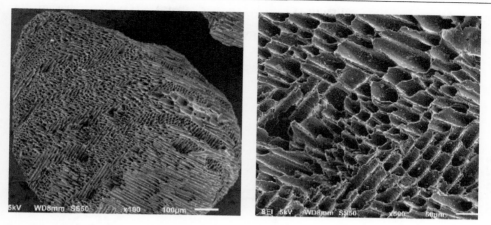

Fig. 6.3 (a) SEM image of wood charcoal (WC 3).

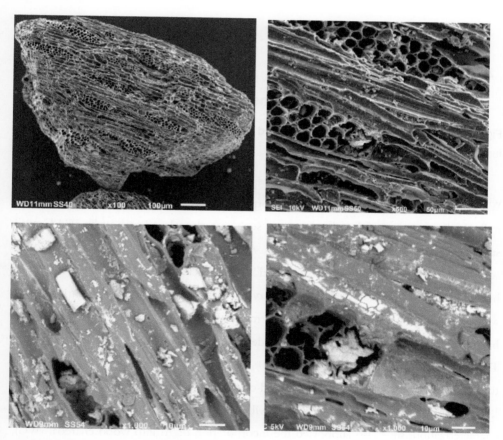

Fig 6.3 (b) SEM images after surface treatment (AFWC 3).

Fig. 6.4 also presents a typical FTIR characterization of wood charcoal before (WC 3) and after the surface functionalization process (AFWC 3).

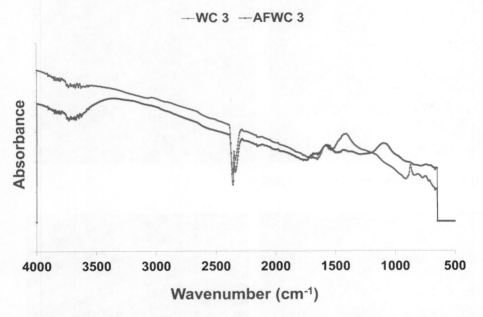

Fig. 6.4 Typical FTIR characterization of wood charcoal before (WC 3) and after the surface fuctionalization process (AFWC 3).

Wood charcoal (WC 3) showed FTIR adsorption bands at $800 - 2000$ cm^{-1}, which may indicate the presence of functional groups such as carbonyl and carboxylic acid $C = O$ streach typical for carbon material surfaces. A very broad band at $2,500- 3,000$ cm^{-1} also suggested the presence of carboxylic acid OH streach. After the surface treatment, however, some adsorption peaks were observed to either disappear, reduce in intensity or were modified. The spectra of the treated material (AFWC 3), showed new adsorption bands at $800 - 1,700$ cm^{-1} which corresponded to Al based OH mode (Frost et al., 1999). The broad adsorption band of WC 3 at $2,500 - 3,000$ cm^{-1}, was significantly modidied, with the appearance of a shaper adsorption band on the AFWC 3 spectra at $3,000 - 3,500$ cm^{-1}, which may be related to the streaching vibrations of Al based OH bonds. The FTIR characterization, thus also supported the incorporation of Al into wood charcoal (WC), with the expected aluminol (AlOH) functionality.

Chapter 6 Aluminol (Al-OH) fuctionalized wood charcoal for treatment of
fluoride-contaminated groundwater: Effect of wood source and field assessment

219

6.3.2 Fluoride removal efficiency of wood charcoal (WC), Aluminol functionalized wood charcoal (AFWC), and effect of the precursor WC

An assessment of the fluoride removal efficiency of untreated WC and AFWC produced from different precursor wood charcoal base materials, based on the batch adsorption experiments is presented in Fig. 6.5 and Table 6.2.

Table 6.2 Fluoride removal efficiency of WC and AFWC.

Adsorbent	WC 3	AFWC 1	AFWC 2	AFWC 3	AFWC 4
F removal (%)	45.7	79.4	95.2	97.4	97.7
Time (h) to attain WHO guideline value of 1.5 mg/L	-	72	2.0	0.5	0.5

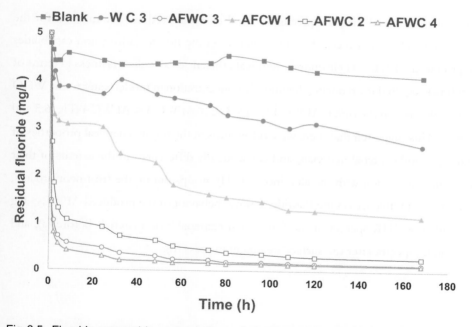

Fig 6.5 Fluoride removal by untreated wood charcoal (WC 3) and aluminum functionalized wood charcoals (AFWC 1, AFWC 2, AFWC 3 and AFWC 4), using different precursor wood charcoal base materials in batch adsorption experiment: Model water: fluoride = 5 ± 0.2 mg/L, HCO_3 = 260 mg/L, pH = 7.0 ± 0.1, adsorbent dose = 10 g/L, shaker speed = 100 rpm.

It was observed that untreated wood charcoal (WC 3) was less effective for fluoride removal (Fig. 6.5). Using an adsorbent dose of 10 mg/L, WC 3 was unable to reduce fluoride concentration of 5 ± 0.2 mg/L in model water to the WHO guideline value of 1.5 mg/L, even after 176 h of contact. This may be attributed to the non metallic nature of the wood charcoal material, a characteristic which is observed to render such carbonaceous materials less effective for the uptake of inorganic species such as fluoride from aquoues solutions (Ayoob and Gupta, 2016). The surface treatment with aluminum, however, significantly enhanced the fluoride removal capabilities of the produced modified materials (AFWC 1, AFWC2, AFWC 3 and AFWC 4), which is presumably due to the presence of aluminol functional groups, as suggested by the FTIR characterization. AFWC 3 was particularly able to reduce the fluoride concentration in model water to the WHO guide line value within 0.5 h of contact.

It was further observed that, the precursor wood charcoal used as base marerial for the surface functionalization process had an influence on the fluoride adsorption capabilities of the produced AFWCs. Their fluoride removal efficiencies as well as fastness in terms of contact times required for reducing the fluoride concentration in model water to the WHO guideline value was in the order: AFWC 1< AFWC 2 < AFWC 3 < AFWC 4 (Fig. 6.5 and Table 6.2). This suggested that there were differences in the physic-chemical properties of the starting wood charcoal materials, and consequently differences in the extents of their surface functionalization with the aluminol (AlOH) groups during the treatment process, and therefore an influence on the fluoride removal behavior of the produced AFWCs. Fig 6.6 presents the FTIR spectra of the four wood charcoal base materials before (a), and corresponding spectra after the surface treatment process (b).

Chapter 6 Aluminol (Al-OH) fuctionalized wood charcoal for treatment of
fluoride-contaminated groundwater: Effect of wood source and field assessment

221

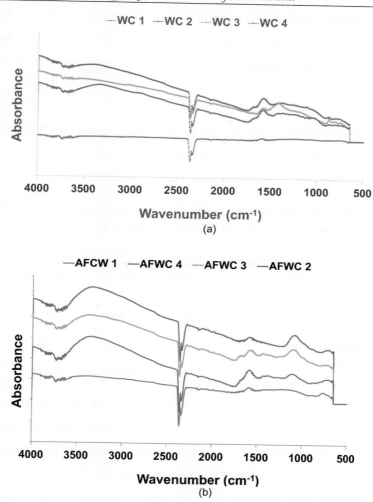

Fig 6.6 FTIR spectra of (a) different precursor wood charcoals (WCs) and (b) spectra of corresponding aluminol fuctionalized wood charcoals (AFWCs).

It was observed from the FTIR spectra of the wood charcoals (Fig. 6.6a) that the intensities of the inherent surface functional groups on their particles, which are responsible for their characterstic surface reactivities, was in the order: WC 1 < WC 2 < WC 3 < WC 4. The corresponding intensities of the Al based functional groups (Fig 6.6 b) after the surface treatment process with Al was also observed to be of the similar trend/order: AFWC 1 < AFWC 2 < AFWC 3 < AFWC 4. These observations suggest that different precursor wood charcoal possessed different intensities of surface fuctional groups (hence differences in surface reactivity), which presumably influenced the extent of the aluminol functionalization   process, and resulting in AFWCs with different fouride removal capabilities.

Table 6.3 presents the BET specific surface areas and porosities for WC 1/AFAC 1 and WC 3/AFWC 3, which were at the extreme ends or most contrasting (i.e low/high) in terms of their fluoride removal efficiencies (Fig. 6.5) and Table 6.2.

Table 6.3 Surface area and porosity of untreated wood charcoal (WC 1 & WC 3) and corresponding surface treated products (AFWC 1 & AFWC 3).

Sample	Weight loss (wt %)	S_{BET} (m^2/g)	V_{pore} (cm^3/g)	V_{micro} (cm^3/g)	S_{meso}
Untreated wood charcoal (WC 1)	8.1	264	0.114	0.106	6
AFWC 1	10.7	269	0.117	0.110	10.7
Untreated wood charcoal (WC 3)	4.9	88	0.060	0.040	10
AFWC 3	10.6	99	0.070	0.440	12

The BET specific area and porosity characterization further revealed significant differences in the physic-chemical properties of the precursor wood charcoal base materials, which was expected to influence both the surface treatment process with Al, as well as the fluoride removal properties of the produced adsorbents. The results obtained in terms of the fluoride adsorption performance of the produced adsorbents (Fig. 6.5 and Table 6.2) could, however, not be explained by difference in surface area and porosity. Even though WC 1 and the corresponding product, AFWC 1, possessed comparatively very good textural properties (i.e surface area and porosity), the adsorbent performed the least with regards to the fluoride removal efficiency (Fig. 6.5 and Table 6.2). Fig. 6.7a and Fig. 6.7b also present the SEM images of WC 1 and that of the corresponding aluminium modified product, AFWC 1, respectively.

Chapter 6 Aluminol (Al-OH) fuctionalized wood charcoal for treatment of fluoride-contaminated groundwater: Effect of wood source and field assessment

223

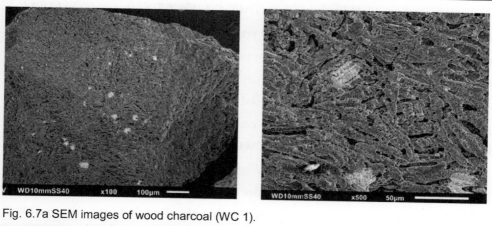

Fig. 6.7a SEM images of wood charcoal (WC 1).

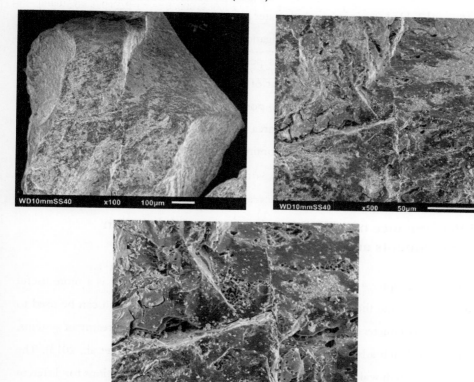

Fig. 6.7b SEM images of aluminium modified product (AFWC 1).

A comparison of the SEM images of WC 1/AFWC 1 (Fig. 6.7) to that of WC 3/AFWC 3 of similar magnifications (Fig 6.3), showed yet another significant difference between the physico-chemical properties of the different precursor wood charcoal base materials with regards to the nature of their pore structures, hence the properties of the produced fluoride adsorbents. The SEM image (Fig. 6.7) showed WC 3 particles as an accumulation of tubes joined to each other, and more structurally ordered, with larger pore size and presumably better interconnected, compared to that of WC 1. Many factors are known to control the adsorption in microporous materials including; the pore size distribution and pore shape (Daley et al., 1996). The nature of the pore structure of WC 3 presumably resulted in AFWC 3 possing more condusive properties for fluoride adsorption compared to AFWC 1, (Fig. 6.5 and Table 6.2), eventhough the later possessed a far higher BET specific surface area. This, however, require further investigation. AFWC 3 was thus selected for further work in continuous flow studies, since its performance, based on the batch screening adsorption experiments, was similar to that of AFWC 4 (Fig. 6.5 and Table 6.2), and in addition the AFWC 3 particles appeared to possess a higher mechanical strength and are more rebust. Given the rather high surface area of AFWC 1, however, a further assessment of the fluoride adsorption capacity under continuous flow conditions also require to be re-visited in the future.

6.3.3 Performance of AFWC in laboratory-scale column experiments and effect of EBCT

As discussed in chapter 4 (section 4.1) a laboratory-scale column study is a more useful approach for assessing the fluoride removal potential of adsorbents, and can be used to generate useful parameters for the design of full-scale adsorption water treatment systems, as compared to batch adsorption studies (Shih, et al, 2003; Quintelas et al., 2013). The results of fluoride removal by AFWC 3 under the continuoues flow conditions for different empty bed contact times (EBCTs) of 6, 12, 18, 24 and 30 min are presented in the form of breakthrough curves (ploted as C_t/C_o versus time) in Fig. 6.8. The effects of EBCT on the fluoride removal performance by AFWC 3 in terms of volume of water treated till the breakthough (1.5 mg/L) are also shown in Table 6.4.

Chapter 6 Aluminol (Al-OH) fuctionalized wood charcoal for treatment of
fluoride-contaminated groundwater: Effect of wood source and field assessment

225

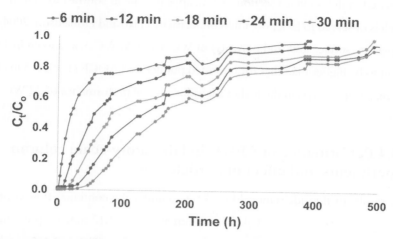

Fig 6.8 Breakthrough curves for fluoride removal by AFWC 3 (particle size 0.8 – 1.12 mm) for different EBCTs: Filtration rate = 5.0 m/h; Model water: fluoride = 5.0 ± 0.2 mg/L, HCO_3^- = 330 mg/L, pH = 7.0 ± 0.2, temp. = 20 °C.

Table 6.4: EBCTs, breakthrough times, mass and volume of adsorbent and volumes of water treated for fluoride removal in AFWC 3 column filter.

EBCT (min)	Breakthrough time (h)	Volume of adsorbent (m^3)	Mass of AFWC 3 (g)	Volume of water treated (m^3)	No. of bed volumes (BV)
6	9	0.0014	381	0.1269	90
12	33	0.0028	765	0.6204	219
18	54	0.0042	1,146	0.7614	180
24	78	0.0057	1,527	1.0857	192
30	104	0.0071	1,911	1.4664	207

It was observed at the time of attaining breaktrough (which corresponded to C_o/C_t = 0.3), that the maximum number of bed volumes (BV) of water treated by AFWC 3 was 219 (Table 6.4). This was found to be 33% higher than the maximum BV for aluminium oxide coated pumice (AOCP), which was 165 BV (Table 4.1 in chapter 4). The superior performance of AFWC 3 over AOCP may be due to differences in the textural properties (surface area and porosity) of the precursor base materials (virgin pumice versus wood charcoal) used in the fluoride adsorbent synthesis process, which most likely influenced their fluoride removal capabilities. The nature of the pore structure of wood charcoal base material (WC 3), and hence influence on the fluoride removal capability of AFWC 3 as discussed in the previous section (6.3.2), appear to be in contrast to that of virgin pumice, which is observed to lack pore interconnecttivity (Tiab and Donaldson, 2004). The textural properties (surface area and porosity) of WC 3 and AFWC 3, as shown in Table 6.3, were found to be higher than that for uncoated pumice and AOCP (Table 3.3 in chapter 3), and this may help to explain the higher fluoride removal performance of AFWC 3 (Cater et al., 1986).

6.3.4 Performance of AFWC in laboratory-scale column experiments and effect of particle size

The results of fluoride removal by AFWC 3 under the continues flow conditions for the two particle size ranges of 0.425 – 0.8 mm and 0.8 – 1.12 mm are presented in Fig. 6.9. As expected, the reduction in particle size range of AFWC 3 (0.425 – 0.8 mm) further improved its fluoride removal performance (Fig.6.9), and the number of bed volumes (BV) of water treated prior to breakthrough (1.5 mg F/L) increased to 282, which represented a further improvement of AFWC performance of 29 %.

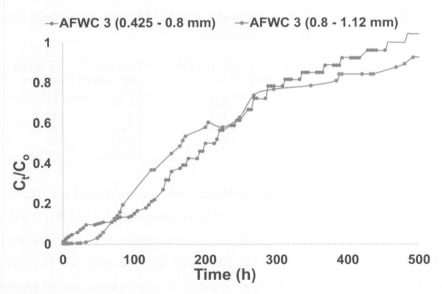

Fig 6.9 Breakthrough curves for fluoride removal by AFWC 3 for particle size ranges: Filtration rate = 5.0 m/h; EBCT = 30 min; Model water: fluoride = 5.0 ± 0.1 mg/L, HCO_3^- = 330 mg/L, pH = 7.0 ± 0.2, temp. = 20 °C.

The fluoride adsorption behaviour of the adsorbent in two particle size ranges may be explained by a number of factors including; the relative importance of two particle-size-dependent processes as well as the relation between particle size and diffusion path length. Compared to the larger particle size, the smaller particle size will have a larger total outer surface area, with more active sites that are readily available for instantanoues fluoride uptake. Thus the dominant process for AFWC 3 with smaller particle size (0.425 – 0.8 mm) is a more rapid external-surface adsorption, particularly in the initial period of the process. A slower adsorption into the interior surface would presumably follow, only after the instantaneous utilization of the most readily available sites on the outer surface. On the other hand, adsorption onto the coarser AFWC 3 (0.8 – 1.12), may be dominated in relative terms more by slow intraparticle adsorption. It further known that the adsorption kinetics or rate at which adsorption equilibrium is established, depends on the diffusion path length. This property (i.e. the diffusion path length) is inturn, a function of many variables including the adsorbent size/diameter. In general, the adsorption kinetics increase as the particle size/diameter decreases, due to a decrease of the diffusion path length. All these factor may account or contribute to the higher performance of the AFWC 3 with a smaller particle size range (Deley et al., 1996; Canales et al., 2013; Salifu et al., 2013). A more rapid fluoride adsorption kinetics by the AFWC 3 of smaller size could be of significant practical importance, as it may result in smaller reactor volumes and ensure higher efficiency and economy (Dawood and Sen, 2012).

6.3.5 Performance of AFWC3 and regenerated AFWC3 (RAFWC 3) for treatment of natural fluoride-contaminated groundwater

The results of field studies for assessing the efficacy of the coarser AFWC 3 (i.e pariticle size range of 0.8 – 1.12 mm, which was available at that time of field work) for treatment of natural fluoride-contaminated groundwater are shown in Fig. 6.10 a & b. Results of an assessment of the field performance of regenerated AFWC 3 (RAFWC 3), and a comparison with the performance of activated alumina (AA) are also included in Fig. 6.10.

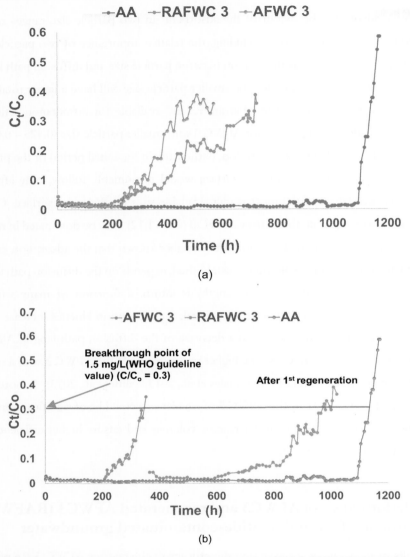

(a)

(b)

Fig.6.10 Breakthrough curves for fluoride removal by (a) AFWC 3, RAFWC 3 and AA, and (b) scenario of comparison taking into consideration a single regeneration of AFWC and the non-regenerable of AA.

AFWC 3 was found capable of reducing fluoride from the measured concentration of 4.88 mg/L in natural groundwater to $\leq$ 1.5 mg/L (WHO guideline value), indicating the laboratory performance was reproducible under field conditions. The number of bed volumes (BV) of water treated before the breakthrough was approximatetly 208. A

Chapter 6 Aluminol (Al-OH) fuctionalized wood charcoal for treatment of
fluoride-contaminated groundwater: Effect of wood source and field assessment

229

comparison of the performance of AFWC 3 with that of AA under similar conditions (Fig. 6.10 a), indicated a higher performance of the later, with 807 bed voulmes of water treated at breakthrough, approximatelt 3.8 times more than AFWC. The particle size range of the AA used for the field assessment was, however, much finer (0.21 – 0.63 mm) than that of AFWC 3 (0.8 – 1.12 mm), which might have contributed to its higher performance (Deley et al., 1996; Canales et al., 2013; Salifu et al., 2013). Moreover the type/grade of AA tested in the field was said to be very effective but not be regenerable, and has to be used only once and disposed off, as mentioned by the supplier. According to Chauhan et al., 2007, however, the use AA for treatement of fluoride-contaminated groundwater in developing countries can only be cost-effective when it can be used for multiple cycles.

An assessment of the performance of regenerated AFWC 3 (RAFWC 3) in the field showed that it was also effective for treating the natural fluoride-contaminated groundwater (Fig 6.10a), therefore indicating its regenerability. RAFWC 3 was found to possess a higher fluoride removal performance than that of AFWC 3, and treated 295 bed volumes of water at breakthrough, representing an increase of 40 % of performance after regeneration. This was therefore similar to the performance of regenerated AOCP (RAOCP). The regenerability of AFWC 3 for reuse could contribute to its cost-effectiveness. Since the AA grade tested in the field is not regenerable and can only be applied once, a comparison of a scenario of the performance of fresh AFWC 3 icluding its 1st cylcle regenerartion to that of AA is presented in Fig. 6.10 (b), which shows AFWC 3's performance as reasonably approaching that of AA. It is, however, expected that AFWC of smaller particle size (0.425 – 0.8 mm) and, it's regenerated version may perform even better under the field conditions. Moreover AFWC 3 exhibits a potential for multiple regeneration, with a likelihood of increase of adsorption capacity with each cycle of regeneration. This, however, require further investigation.

6.3.6 Acidity of AFWC 3 and RAFWC 3 surfaces

In section 4.5.2 of chapter 4, an attempt was made to explain the factors contributing to the increased fluoride removal performance of AOCP after regeneration when it is fully exhausted, which is also similar to the observations for AFWC 3, after regeneration. In this chapter, the acidity of AFWC 3 and RAFWC 3 surfaces obtained from potentiometric titration techniques were additionally compared in an attempt to get a further insight into their aptitude to adsorb fluoride ions from aqueous solutions. The protonation of the surface aluminol fuctional groups (AlOH) on the adsorbents to form reactive sites play a critical role in the fluoride adsorption process. The protonated forms ($\equiv AlH_2^+$) acts as acid sites for the fluoride adsorption. The protonation reactions of the two adsorbents surfaces were modelled using the 1 site 2 pKa model (Turner and Fein, 2006), from which the protonation constants (pKa) were determined. In this approach, the AFWC 3 and RAFWC 3 surfaces were considered to have one type of amphoteric surface functional group that is the aluminol (AlOH) group, whereby the following two protonation reactions can be written with equilibrium constants K_1 and K_2, respectively:

$$\equiv AlOH^0 = \equiv AlO^- + H^+ \tag{6.1}$$

$$\equiv AlH_2^+ = \equiv AlOH^0 + H^+ \tag{6.2}$$

In equation (2), the $\equiv AlOH^0$ functional groups exhibit basic behavior, and adsorbing protons to form positively charges surface species (Turner and Fein, 2006), i.e. the conjugate acid/protonated form ($\equiv AlH_2^+$). The potentiometric titration data were analyzed using the ProtoFit software to obtain the pKa values for the AFWC 3 and RAFW3, for assessing and comparing the strengths of their surface acidities. The smaller the pKa value for a given surface the stronger the acidity.

The pKa determined from the potentiometric data analysis for RAFWC 3 was -9, which was slightly less than the pKa for AFWC 3 = -8.9. This suggested the RAFWC 3 surfaces were slightly more acidic/reactive for fluoride adsorption compared to the AFWC 3 surfaces, which may further to help explain the higher fluoride removal performance of the former (i.e. RAFWC 3). Thus a combination of increased amount of surface species created in the regeneration/re-coating process (section 4.5.2), coupled with an enhanced surface acidity/reactivity of the regenerated material may be responsible for the enhanced performance. The enhaced surface acidity of RAFWC 3 is presumably due to the incorporation of fluoirine atom in the structure of the aluminol functional of RAFWC 3,

Chapter 6 Aluminol (Al-OH) fuctionalized wood charcoal for treatment of fluoride-contaminated groundwater: Effect of wood source and field assessment

231

whereby F$^-$ replaces some of the hydroxyl (OH)/oxygen group in the function group, which changes its surface structure and chemistry. Because fluorine is more electronegative (electronegativity = 4.0) than the hydroxyl/ oxygen (electronegativity = 3.5), electron density is pulled away from the adjacent hydroxyl groups (i.e. the remaining OH groups coordinated to Al) towards the fluoride atom. This weakens the O-H bond of the adjacent hydroxyl groups and makes their hydrogen more acidic, hence more subceptible for nucleuphilic interaction with fluoride in solution compared to the interactions with the functional groups on the AFWC 3 surface (Ghosh and Kydd, 1985).

6.3.7 Mechanism of fluoride removal onto AFWC 3

Fig. 6.11 show the FTIR spectra for AFWC 3 (a) before and (b) after fluoride adsorption. The spectra before fluoride adoption (Fig. 6.11a) exhibited adsorption peaks in the low and high frequency regions, consistent with the presence of aluminum based functional groups. The adsorption peaks were, however, significantly modified including; reduction in intensities, and/or disappearance of some peaks after fluoride adsorption, and the appearance of new peaks in the region of 500 – 800 cm^{-1} (Fig.6.11b) The new peaks in the region of 500 – 800 cm^{-1} are consistent with the Al – F stretching vibrations, which can be attributed to the complexation of fluride ions with Al (Kumar et al., 2011).

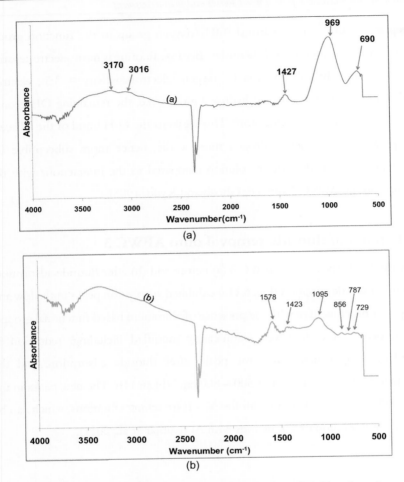

Fig. 6.11 FTIR spectra of AFWC 3 (a) before and (b) after fluoride adsorption.

6.3.8 Stability of Al incorporated into wood charcoal for aluminol (AlOH) functionality

Fig. 6.12 presents results of the leaching experiment for examining the stability of Al incorporated into AFWC, which shows amount of Al leached from the adsorbent as function of pH, after 24 h of contact.

The amount of Al leached from AFWC within the experimental pH range of 5 – 9, ranged between 0.004 mg/L and 0.057 mg/L. The amount of Al leached at a neutral pH of 7, in particular, was 0.004 mg/L (Fig. 6.12), which is by far less than the old WHO guideline

Chapter 6 Aluminol (Al-OH) fuctionalized wood charcoal for treatment of fluoride-contaminated groundwater: Effect of wood source and field assessment

233

value of 0.2 mg/L that is incorporated in most of national drinking water standards. This suggest that the use AFWC for water defluoridation may not contaminate the treated water with excess Al in practice.

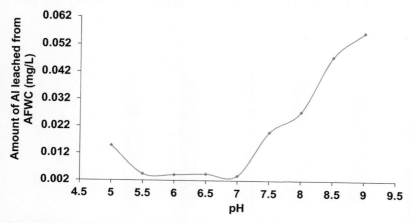

Fig. 6.12 Leaching of Al from AFWC under different pH conditions. Leaching solution = 50 mL of demineralized water

6.3.9 Effects of co-ions, pH and storage time on the fluoride adsorption performace of AFWC

Since aluminol functionalized wood charcoal (AFWC) appear to be the most promising adsorbent among the 3 developed water defluoridation materials, with a potential for field application, the effects of co-ions, pH and long-term storage on its fluoride removal performance was further studied. The results of the batch adsorption experiment on the effect of co-ion are presented in Fig 6.13 (a & b).

Similar to GACB, and at the same co-ion concentrations of 2.5 mM, phosphate competed most with the fluoride ions for adsorption sites on AFWC 3 (Fig. 6.13 a). The order of competition of co-ions with the fluoride ions was also similar as follows: phosphate > bicarbonate ≈ sulphate ≈ nitrate > chloride. Under similar conditions, however, the reduction of the fluoride adsorption by AFWC 3 in the presence of phosphate was far less, and limited to only 5 %, whereas the reduction of the adsorption onto GACB was 39 % (Fig.5.9). This indicated AFWC 3 has a superior performance over GACB. Further adsorption experiments conducted using lower concentrations of phosphate (= 0.01 mM) and nitrate (= 1.6 mM) that can be found in groundwater, showed no impact at all of these co-ions on fluoride adsorption onto AFWC (6.13 b), whereas they had a slight impact on fluoride adsorption onto GACB. This indicated the potential of AFWC for field application.

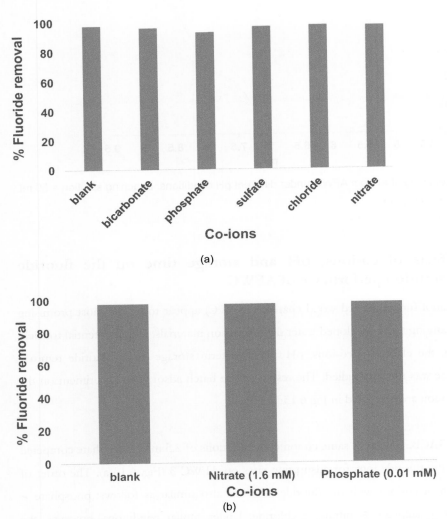

(a)

(b)

Fig. 6.13 Effects of co-anions on fluoride removal by AFWC, (a) co – ion: 2.5 mM and (b) co-ion concentrations for nitrate and phosphate that can be found in groundwater. Model water: Fluoride = 5 ± 0.2 mg/L, pH = 7.0 ± 0.1, Contact time = 172 h, Temp. = 20ºC, AFWC dose = 10g/L.

Chapter 6 Aluminol (Al-OH) fuctionalized wood charcoal for treatment of
fluoride-contaminated groundwater: Effect of wood source and field assessment

235

The effect of pH in the range of 3 – 11, on the adsorption of fluoride onto AFWC is also presented as Fig 6.14. AFWC showed good fluoride removal in the pH range of 6 – 9, which therefore makes it suitable for groundwater treatment without need for pH adjustment.

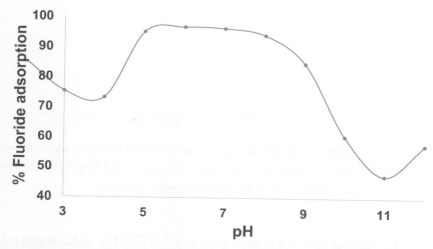

Fig. 6.14 Effect of pH on fluoride adsorption onto AFWC. Model water: Fluoride = 5 ± 0.2 mg/L, Contact time = 172 h, Temp. = 20°C, AFWC dose = 10g/L.

The effect of storage time under normal room conditions, on the fluoride removal performance of AFWC is presented in Fig. 6.15. Unlike granular aluminium coated bauxite (GACB), the fluoride removal capability of AFWC remained the same after 8 months of storage. This suggested the Al incorporated into the wood charcoal matrix in the form of oxides did ont undergo any transformations that could affect the fluoride adsorption properties. The effects of longer storage times on the performance of AFWC, however, required further investigation, since in practice adsorbents could be in storage for much longer times than 8 months, before application.

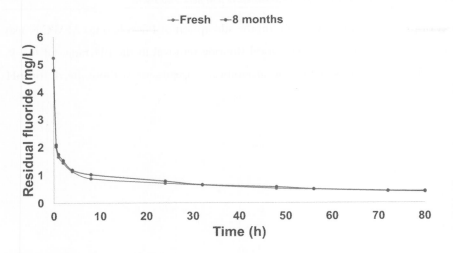

Fig. 6.15 Effect of storage time on fluoride removal by aluminum functionalized wood charcoal (AFWC) in batch adsorption experiments. Model water: fluoride = 5 ± 0.2 mg/L, HCO_3 = 260 mg/L, pH = 7.0 ± 0.1, adsorbent dose = 10 g/L, shaker speed = 100 rpm.

6.3.10   Leaching test for waste (spent) AFWC for safe disposal

Given that AFWC shows a better potential to be applied as a fluoride adsorbent in the field in practice, a study on its safe disposal into the environment, when exhausted was additionaly conducted. The US-EPA's toxicity characterization leaching procedure, as described in section 4.2.5 (and in detail elsewhere (US-EPA, 1992)), was used to characterize the leaching properties of both spent AFWC that was treated with Al for fluoride stabilization, and spent AFWC without any treatment, to determine if the spent AFWC has the characteristics of hazardous materials or otherwise. The results of the leaching test are presented in Table 6.5.

Table 6.5 TCLP extraxts from spent AFWC.

Spent AFWC	Average fluoride concentration (mg/L) in triplicate samples
Untreated (without fluoride stabilazation)	10.71 ± 0.25
Treated with Al (fluoride stabilized)	1.16 ± 0.03
NIPDWS regulatory level at 32ºC	140

Chapter 6 Aluminol (Al-OH) fuctionalized wood charcoal for treatment of
fluoride-contaminated groundwater: Effect of wood source and field assessment

237

Similar to spent aluminium oxide coated pumice (AOCP), the fluoride concentrations of the extracts from spent AFWC (with and without stabilazation) were found to be far less than the threshold established by US-EPA, (i.e. 100 times the National Interim Primary Drinking Water (NIPDW) standard for fluoride). This showed the non-hazardous nature of spent AFWC, and, hence could be disposed of in a simple landfill without risk of environmental pollution.

6.4 Conclusions

The naturally higher textural properties (i.e surface area and porosity) of locally available wood charcoal makes it a more suitable base material for the sysnthesis of an aluminum modified fluoride adsorbent, compared to base materials such as bauxite and pumice.

Different wood charcoal produced from different trees/plants exhibited different physic-chemical properties. These included differences in intensities of their inherent surface functional groups, BET surface areas, porosities and nature of the pore structures. The source of wood charcoal used as precursor base material for the production of aluminol functionalized wood charcoal (AFWC), has an influence on its fluoride removal capabilities.

The particle size range of AFWC has an effect on its fluoride adsorption performance, with smaller particle size performing better (i.e. 29 %) in terms of number of bed volume of water treated before the breakthrough.

Similar to AOCP, the fluoride removal capability of AFWC was increased by 40 % after regeneration.

AFWC and its regenerated version (RAFWC) were found capable of reducing fluoride concentraton of 4.88 mg/L in natural fluoride-contaminated ground water to ≤ 1.5 mg/L under field conditions.

The grade of activated alumina (AA) tested in the field performed about 3.8 times better than the AFWC tested. The particular type of AA used in the field study may likely not, however, be cost-effective for application in developing countries since, in contrast to AFWC, it can not be regenerated and has to be used once and thrown away.

The field performance of AFWC and RAFWC were found encouraging, though the AA have shown, under the field test conditions, superior performance, having in mind that the AA grade tested may be relatively expensive and cannot be regenated, while further optimization and multiple regeneration of AFWC are likely possible. The current fluoride removal capacity of AFWC, however, require to be improved.

Co-ions at concentrations commonly found in groundwater has no impact on fluoride uptake by AFWC, which makes the adsorbent potentially suitable for groundwater defluoridation in practice.

AFWC exhibited good fluoride removal in the pH range of 6 – 9, which also makes it suitable for the treatment of fluoride-contaminated groundwater without need for pH adjustment.

Spent AFWC (with or without fluoride stabilazation) has characteristics of non-hazardous material and can be disposed of safely into the environment in a simple land-fill, when it can no longer be used.

References

Ahmed I. Abdelrahman, A. I., Thickett, S.C., Liang, Y., Ornatsky, O., Baranov, V., and Winnik, M.A., 2011. Surface functionalization methods to enhance bioconjugat ion in metal-labeled polystyrene particles. Macromolecules, 44, 4801 – 4813.

Chapter 6 Aluminol (Al-OH) fuctionalized wood charcoal for treatment of
fluoride-contaminated groundwater: Effect of wood source and field assessment

239

Alemdaroglu, T., 2001. Determination methods for the acidity of solid surfaces. Comm. Fac. Sci. Uni. Ank. Series B, 47, 27-35.

Beguin, F., and Frackowiak, E., 2010. Carbon for electrochemical energy storage and conservation systems. CRC Press, Taylor and Francis group, New York.

Bhatnagar, A., Kumar, E., Sillanpää, M., 2011. Fluoride removal from water by adsorption—A review. Chem. Eng. J., 171, 811– 840.

Canales, M.R., Huade Guan, H., Bestland, E., Hutson, J., Simmons, C.T., 2013. Particle-size effects on dissolved arsenic adsorption to an Australian laterite Environ Earth Sci. 68, 2301–2312.

Cater, D.L., Mortland, M.M., Kemper, W.D., 1986. Speci fi c surface. In: Part 1.Physical and Mineralogical Methods- Agronomy Monographs no.9, second ed. American Society of Agronomye Soil Science Society of America, Madison, pp. 413 - 422.

Cecen, F., and Aktas, O., 2011. Activated Carbon for Water and Wastewater treatment: Historical perspective of activated carbon adsorption and its intergration with biological processes.WILEY-VCH Verlag GmbH & Co. KGaA, Weinheim, Germany.

Chauhan, V.S., Dwivedi, P.K., Leela Iyengar, L., 2007.Investigations on activated alumina based domestic defluoridation units, J. of Hazard. Mater. B139, 103–107.

Czernik, S., 2008. Fundamentals of Charcoal Production, National Bioenergy Center, IBI Conference on Biochar, Sustainability and Security in a Changing Climate. September 8-10, Newcastle, U.K.

Daley, M.A., Mangun, C.L., and J. Economy, J., 1996. Predicting adsorption properties for ACFs. Preprints of Div. of Fuel Chem., 41(1), 326-330.

Dawood, S., Sen, T.K., 2012. Removal of anionic dye Congo red from aqueous solution by raw pine and acid-treated pine cone powder as adsorbent: equilibrium, thermodynamics, kinetics, mechanism and process design, Water Res. 46, 1933–1946.

Faust, S.D., Aly, O.M.   Chemistry of water treatment. 2nd ed, CRC Press LLC, Florida, USA, 2008.

Frost, R.L., Kloprogge, J.T., Russel, S. C., Szetu, J.L., 1999. Vibrational Spectroscopy and Dehydroxylation of Aluminum (Oxo) hydroxides: Gibbsite. Applied Spectros. 53 (4), 423 -434.

Ghosh, A.K and Kydd, R. A., 1985. Fluorine-promoted catalyst. Cataly. Rev. Sci. Eng. 27(4), 539 – 589.

Gupta, A.K., and Ayoob, S., 2016. Fluoride in drinking water: status, issues and solutions. CRC Press, Taylor and Francis group, New York.

Figueiredo, J., 2013. Functionalization of porous carbons for catalytic applications. J. Mater. Chem. A., 1, 9351- 9364.

Kumar, E., Bhatnagar, A., Kumar, U., Sillan, M., 2011. Defluoridation from aqueous solutions by nano-alumina: Characterization and sorption studies, J. of Hazard. Mater. 186, 1042–1049.

Kumar, E., Bhatnagara, A., Kumar, U., Sillanpaa, M., 2011. Defluoridation from aqueous solutions by nano-alumina: Characterization and sorption studies. J. of Hazard. Mater.186, 1042-49.

Ma, W., Yah, W.O., Otsuka, H., and Takahara, A., 2012. Surface functionalization of aluminosilicate nanotubes with organic molecules. Beilstein J. Nanotechnol, 3, 82-100.

Chapter 6 Aluminol (Al-OH) fuctionalized wood charcoal for treatment of
fluoride-contaminated groundwater: Effect of wood source and field assessment

241

Muller, B.R., 2010. Effect of particle size and surface area on the adsorption of albumin-bonded bilirubin on activated carbon. CARBON 48, 3607 – 3615

Nishimiya, K., Hata, T., Imamura, Y., Ishihara, S., 1998. Analysis of chemical structure of wood charcoal by X-ray photoelectron spectroscopy. J. Wood Sci., 44, 56 – 61.

Pastor-Villegas, J., Pastor-Valle, J.K,. Meneses Rodrı´guez, J.M., Garcıa Garcıa, M., 2006. Study of commercial wood charcoals for the preparation of carbon adsorbents. J. Anal. Appl. Pyrolysis, 76, 103-108.

Quintelas,C., Pereira, R., Kaplan, E., Teresa Tavares, T., 2013. Removal of Ni (II) from aqueous solutions by an Arthrobacter viscosus biofilm supported on zeolite: From laboratory to pilot scale. Bioresour. Technol 142, 368–374.

Reitzel, K., Jensen, H.S., Egemose, S., 2013. pH dependent dissolution of sediment aluminium in six Danish lakes treated with aluminium. Water Res., 4 7, 1409- 1420.

Shen, W., Li, Z., and Liu, Y., 2008. Surface Chemical Functional Groups Modification of Porous Carbon Recent Patents on Chem. Eng., 1, 27-40.

Shih, T.C., Wangpaichitr, M., Suffet, M., 2003. Evaluation of granular activated carbon technology for the removal of methyl tertiary butyl ether (MTBE) from drinking water. Water.Res.37, 375-385.

Thakur, V. K., Thakur, M.K (Ed.) .2016. Chemical functionalization of carbon nanomaterials: chemistry and applications. CRC Press, Taylor and Francis group, New York.

Thomas, W.J., Crittenden, B., 1998 Adsorption Technology & Design. Reed Educational and Professional Publishing, Woburn, U.K.

Tiab, D and Donaldson, E.C., 2004. Petrophysics: Theory and practice of measuring reservoir rocks and fluid transport properties. Elservier, London.

Turner, B.F., Fein, J.B., 2006. Protofit: A program for determining surface protonation constants from titration data, Computers & Geosciences. 32, 1344 – 1356.

US-EPA. 1992. Toxicity Characteristic Leaching Procedure (TCLP), U.S, EPA. Test methods for evaluating solid waste, Physical/Chemical methods, SW-846.

Yin, C. Y., Aroua, M., and Daud, W., 2007. Review of modifications of activated carbon for enhancing contaminant uptakes from aqueous solutions. Sep. and Purif. Techn., 52, 403-415.

7

General Conclusions

7.1 Overall conclusions and perspective

Groundwater remains the most important source for community/rural water supply in the Northern region of Ghana. One of the main challenges with regards to its utilization, however, is to get an insight into the occurrence, genesis and distribution of excess fluoride in the groundwater, particularly in the eastern corridor of the region (where the problem is more prominent), and to use this knowledge to help in planning for the provision of safe drinking water for those high fluoritic areas.

Because of the permanent risk and the lack of known effective treatment for fluorosis, treatment of fluoride-contaminated water is a necessity, to avoid ingestion of excess fluoride, as a preventive measure. Among the technological options for treating fluoride-contaminated groundwater, adsorption technique is considered appropriate, however, a main challenge with the application of adsorption is the availability of a suitable adsorbent. Additionally, a related challenge associated with the use of any given fluoride adsorbent is its potential for regeneration that will allow re-use, as that is necessary for its practical and economic viability. The safe disposal of a fluoride-saturated adsorbent, (when it can no longer be re-used), in order to prevent environmental/groundwater contamination, also remains a challenge.

There are also significant challenges associated with the development of a fluoride adsorbent capable of treating natural fluoride-contaminated groundwater under field condition, after it has been synthesized/produced and tested in the laboratory. Fluoride removal performance under laboratory conditions may not necessarily be re-producible under field conditions.

Based on the research described in this thesis, it can be concluded that, aside from the boreholes with fluoride concentration either beyond 1.5 mg/L or less than 0.5 mg/L (WHO guideline values), groundwater in the study area (based on the parameters analyzed), is generally chemically acceptable and suitable for domestic use. It was found that the predominant mechanisms controlling the enrichment of fluoride in the study area include; calcite precipitation and Na/Ca exchange processes, both of which deplete Ca from the groundwater, and promote the dissolution of fluorite, hence increase of fluoride concentration in the groundwater system. The mechanisms also include F^-/OH^- anion exchange processes, as well as evapotranspiration processes which concentrate fluoride ions, thus also increasing its concentration. Developed maps showing the spacial distribution of fluoride (both known and predicted) in areas where fluoride presence in groundwater was analyzed, and areas where no water samples were analyzed, respectively, could be useful for water service providers in the planning of strategies for the provision of safe drinking water in the fluoritic areas.

This research also clearly demonstrated that modifying the particle surfaces of indigenous/locally available materials (pumice, bauxite and wood charcoal), by an Al coating/incorporation process, by exploring the hard soft acid base (HSAB) concept, was a useful approach to develop innovative fluoride adsorbents. The Al-modified local materials were capable/effective in reducing fluoride concentration of 5 ± 0.2 mg/L in model water to ≤ 1.5 mg/L, under both laboratory batch and continuous flow conditions. Al-modified wood charcoal was found to be most effective and promising, with a generally good potential for application for the treatment of fluoride-contaminated groundwater in fluoritic areas of developing countries, where charcoal is available. Kinetic and isotherm analyses, pHp.z.c measurements, FTIR and Raman spectroscopic analyses including thermodynamic calculations, revealed that the mechanism of fluoride adsorption onto the Al-modified adsorbent materials was complex and, involved both physisorption and chemisorption processes. The fluoride adsorption capacities of the produced adsorbents described in this thesis were found to be either comparable or higher than the capacities of some fluoride adsorbents reported in literature. It was also found that the produced adsorbents could be regenerated when exhausted, which could contribute to their economic viability in practical applications. In contrast to other fluoride adsorbents, it was further observed that the adsorption capacities of the adsorbents described in this thesis were not only 100 % restored after the 1st cycle of regeneration, but increased by more than 30%.

Based on the US-EPA's Toxicity characterization leaching procedure (TCLP), the spent adsorbent (i.e. when the produced adsorbent is exhausted, and can no longer be re-used) was found to be non-hazardous and could be disposed of safely into the environment using a simple landfill.

This research also clearly demonstrated that the Al-modified indigenous media that were produced, including the regenerated ones, were capable and effective in reducing fluoride concentration of 4.88 mg/L in natural fluoride-contaminated groundwater to ≤ 1.5 mg/L under field conditions. This showed that the results of fluoride removal performance obtained in the laboratory were also reproducible under real field conditions.

The comparability of the fluoride removal performance of the Al-modified indigenous media with that of activated alumina (AA), the industrial standard, was found to depend on the type of AA (i.e. the grade of AA) with regards to its particle size range, and its regenerability (or otherwise), when exhausted. A comparison under similar laboratory conditions, of the fluoride removal performance of Al-modified pumice (i.e. AOCP) with that of an AA grade of similar particle size range (0.8 – 1.12 mm) and which was also described by the supplier as regenerable, revealed that the kinetics of adsorption by the AOCP was faster than that of the AA. A comparison under similar field conditions of the performance of Al-modified wood charcoal (i.e. AFWC 3, which was found to be the most promising among the developed adsorbents), with the performance of a different AA grade, with much finer particle size (0.21 – 0.63 mm), described by the suppler as very effective but not regenerable, showed that the later (i.e. AA) possessed a much higher fluoride removal capacity than that of the AFWC 3. It was concluded, based on available literature that the finer particle size range of the AA grade used for the field study, might have contributed to its superior performance. Based on literature, however, it was revealed that an AA grade that is not regenerable may not be cost-effective for application in developing countries. It was further concluded that the field performance of Al modified wood charcoal (AFWC 3), the most effective adsorbent developed in this study, and its regenerated version, demonstrated encouraging fluoride adsorption capacities, and it is too early to conclude the superiority of the performance of AA over that of the modified wood charcoal since the former (i.e. AA) cannot be regenerated. On the other hand the modified wood charcoal showed good potential for multiple regeneration, which could further improve the economic and pratical viability of the adsorbent. Spent AFWC 3 was also found to be non-hazardous, and could be disposed of in a simple landfill, which could possible reduce the cost of handling and disposal, and presumably contribute to lowering operational cost. The current fluoride adsorption capacity of AFWC 3, however, requires further improvement.

The findings and conclusions of this component of the research described the development of alternative innovative adsorbents for the treatment of fluoride-contaminated groundwater. The information provided may be helpful for water service providers including government, non-governmental organizations (NGOs) and engineers in search of a potential adsorbent for the treatment of fluoride-contaminated groundwater, especially in developing countries.

The remainder of this concluding chapter discusses each individual research aim in more detail in separate sections, in order to place the objectives in a broader perspective. Finally, a general outlook with recommendations for further research are given.

7.2 Fluoride occurrence in groundwater in the Northern region of Ghana

The aim of the work in this section was to study the groundwater chemistry in the Northern region of Ghana, with a focus on the occurrence, distribution and genesis of high fluoride waters in the eastern corridor of the region, where the problem of excess fluoride in groundwater was thought to be most prominent.

Three hundred and fifty seven (357) groundwater samples taken from boreholes drilled in the study area, were analyzed for the chemical data using standard methods. Piper graphical classification, univariate analysis, Pearson's correlation, principal component analysis (PCA) and thermodynamic calculations were used as an approach to gain insight into the groundwater chemical composition and the dominant mechanisms influencing the occurrence of high fluoride waters. The spatial distribution of fluoride in the study area was examined by superimposing the geo-referenced groundwater chemical data onto a digitized geology map of the area. Inverse distance weighting interpolation (IDW) was also applied to the spatial dataset to produce a prediction map for fluoride for the study area.

Twenty three percent (23%) of 357 groundwater samples from the eastern corridor of the Northern region of Ghana, for which the physico-chemical parameters were studied were found to have fluoride concentrations exceeding 1.5 mg/L, the WHO guideline for drinking water, with concentrations as high as 11.6 mg/L. Human consumption of water from these wells can result in the incidence of fluorosis. Fifty two percent (52%) of the groundwater sample were within the acceptable fluoride concentration range of 0.5-1.5 mg/L, while twenty five percent (25%) were found to have fluoride concentrations below 0.5 mg/L, which makes the population using those wells prone to dental caries. Results of the hydrochemical analysis showed that aside from the boreholes with either elevated concentrations of fluoride (beyond 1.5 mg/L) or low concentration ($<$ 0.5 mg/L), groundwater in the study area, based on the parameters analyzed, is chemically acceptable and suitable for domestic use, given that other parameters not covered in the study are also within acceptable limits.

The geochemical processes that control the overall groundwater chemistry in the study area possibly also promote fluoride enrichment of the groundwater. These processes include mineral dissolution reactions, ion exchange processes and evapotranspiration. The predominant mechanisms controlling the fluoride enrichment probably include calcite precipitation and Na/Ca exchange processes, both of which deplete Ca from the groundwater and promote the dissolution of fluorite. The mechanisms also include F^-/OH^- anion exchange processes, as well as evapotranspiration processes which concentrate fluoride ions and increase its concentration. The spatial mapping showed that high fluoride levels in the eastern corridor (the study area) occur predominantly in the Saboba and Cheriponi districts, and also in the Yendi, Nanumba North and South districts of Northern region of Ghana.

7.3 Drinking water defluoridation using aluminum (hydr)oxide coated pumice: Synthesis, equilibrium, kinetics and mechanism

The toxic effects of fluoride on human health, when consumed in excess amounts for long periods is well known. There are several techniques for purifying water with excess fluoride before consumption. Among these techniques, the adsorption process is generally considered the most appropriate, provided a suitable adsorbent is available. Literature review suggested that several studied adsorbents have shown certain degrees of fluoride adsorption capacity, however, applicability of most is limited. This is either due to lack of socio-cultural acceptance, the general lack of sustainability due to non-amenability for local production of industrial standard fluoride adsorbents, high cost, or effectiveness only in extreme pH conditions. There is therefore continuous interest to search for alternative adsorbents. The use of modified locally available materials as fluoride adsorbent also attracts the interest of researchers, as that could possibly allow local production that could contribute to both reduced production costs as well as long-term sustainability.

The aim of this study was therefore to modify the particle surfaces of an indigenous material, pumice, by an aluminum coating process in order to create hard surface sites for fluoride adsorption from drinking water, whereby the hard soft acid base (HSAB) concept was explored.

Al modified pumice, AOCP, was found effective in reducing fluoride in model water from 5.0 $\pm$0.2 mg/L to $\leq$ 1.5 mg/L (WHO guideline value). An attempt to improve on the fluoride adsorption efficiency of AOCP, through a thermal treatment process rather negatively affected the removal efficiency, contrary to expectations. This therefore suggested that thermal treatment of AOCP in its synthesis process may not augment for fluoride removal and should be avoided. AOCP was found to exhibit good fluoride adsorption efficiency within the pH range 6-9, which makes it potentially suitable for groundwater treatment, without any need for pH adjustment with the associated cost and operational difficulties, especially in remote areas of developing countries. Based on batch kinetic experiments, the results from this study also revealed that at a neutral pH of 7.0 $\pm$

0.1, which is a more suitable condition for groundwater treatment, the kinetics of fluoride adsorption by AOCP were comparable or perhaps faster in the initial period of contact than that of activated alumina (AA), of similar particle size range (0.8 -1.12 mm). AA is considered the industrial standard fluoride adsorbent. It was thus concluded that AOCP is promising and could possibly be a useful fluoride adsorbent in developing countries, where indigenous materials (i.e. pumice) could be used, resulting in potentially cheaper adsorbent.

The outcome of this study may be useful for water service providers/engineers in search of fluoride adsorbents that could possibly be produced locally, especially in regions where the virgin pumice material is locally available.

7.4 Laboratory-scale column filter studies for fluoride removal with aluminum (hydr)oxide coated pumice: Filter runs with freshly synthesized and regenerated adsorbent and options for disposal of fluoride-saturated adsorbent

As described in the previous section, an assessment of the efficacy of laboratory synthesized aluminum oxide coated pumice (AOCP) for fluoride removal in batch adsorption experiments, demonstrated that the material was capable of defluoridating model water with excess fluoride to $\leq$ 1.5 mg/L. Even though batch adsorption experiments are relatively quick and easy methods of assessing fluoride removal performance of adsorbents, the outcome of such experiments are not as informative as that from continuous-flow column experiments. Unlike batch adsorption experiments, laboratory-scale column experiments are capable of simulating the dynamics of a fixed bed reactor that can be used to generate parameters for optimizing the design of full-scale adsorption-based water treatment systems. Additionally, the regenerability of ACOP when exhausted is a factor that may contribute to its practical and economic viability. Furthermore the ease (or not) of disposing spent (fluoride-saturated) AOCP, safely without any environmental pollution is a factor that may contribute to its suitability for application in practice.

The aims of this section of the research study were therefore to; (i) assess the fluoride removal behavior of AOCP under dynamic conditions using a laboratory scale fixed-bed adsorption column, (ii) investigate the effects of empty bed contact time (EBCT) on the performance of AOCP, and determine the contact time for optimal use of the adsorbent capacity, (iii) explore a simple method for regeneration of exhausted AOCP, and, (iv) explore a simple treatment method for stabilizing waste (fluoride saturated) AOCP to reduce the risk of environmental pollution with fluoride, when disposed.

In addition to its efficacy for fluoride adsorption under batch conditions, the results of this part of the study established the capability of AOCP of reducing fluoride concentration from 5 ± 0.2 mg/L to ≤ 1.5 mg/L under continuous flow conditions. An empty bed contact time (EBCT) of 24 min was found a suitable guide for design of full-scale water treatment systems that would allow optimal use of the AOCP fluoride adsorption capacity.

The study also established that AOCP could be regenerated for re-use, when exhausted. In contrast to the properties of other adsorbents, it was further found in this study that the fluoride adsorption capacity of exhausted AOCP, after the first cycle of regeneration, was not only fully (100%) restored, but *increased* by more than 30% under batch conditions, and more than 50% under continuous flow conditions. This therefore suggested the effectiveness of the regeneration approach, which has a potential to contribute to the practical and economic viability of AOCP as water defluoridation adsorbent. However further research is required to fully understand the phenomena occurring during regeneration.

Unlike many other fluoride-saturated adsorbents, a characterization of the waste (fluoride saturated) AOCP (with or without stabilization/treatment), based on the US-EPA toxicity leaching characterization procedure (TCLP), revealed that the waste material was non-hazardous, and could be disposed of safely into a simple landfill without risk of environmental/groundwater re-contamination. This can reduce the cost of handling and disposing of spent AOCP media.

Based on Pearson's Chi-squared (χ^2) statistics, it was found that the Adam-Bohart model predicted very well the initial part of the of the breakthrough curves for the fluoride–AOCP system, while the full breakthrough curve could be adequately descried by the Thomas and BDST models. The parameters determined for these models may be helpful to engineers interested in optimizing the design of AOCP column filters for ground water defluoridation, with reduced laboratory experimentation, which could save time and cost. The developed models can also be useful for water treatment plant operators, for prediction of breakthrough times when the filters are in operation, thus determining the times when adsorbent regeneration or replacement is required.

7.5   Fluoride removal from drinking water using granular aluminum-coated bauxite as adsorbent: Optimization of synthesis process conditions and equilibrium study

With regards to the interest in modifying the physico-chemical properties of locally available materials for groundwater defluoridation, bauxite is one such materials that is indigenous in many countries, which could possibly be used as a base material for the synthesis of a fluoride adsorbent, especially in regions where pumice may not be locally available. Raw bauxite has known defluoridation properties, however, the fluoride adsorption capacity is limited and, modifications of its physico-chemical properties for improved performance appear not have been well studied. Moreover, some studied and reported defluoridation materials are either of fine particles or powder that could make separation from aqueous solution difficult. Such materials could also cause clogging and/or low hydraulic conductivities when applied in fixed-bed adsorption systems. Raw bauxite on the other hand is robust and possess sufficient mechanical strength, and could be used as a base material for the production of a granular adsorbent with enhanced defluoridation capability that can also overcome limitations such as clogging and/or low hydraulic conductivities in fixed-bed adsorption systems.

The properties of an adsorbent material, which may contribute to the number of available active sites for fluoride uptake may include both the surface area, as well as the nature of the surface (i.e. its reactivity/affinity for fluoride ions). Therefore in the process of developing a fluoride adsorbent material, methods or process conditions that enhance both the surface area and reactivity/affinity of the produced adsorbent, require attention.

The aim of this component of the research study was therefore; to explore the possibility of synthesizing granular aluminum coated bauxite (GACB) as a novel fluoride adsorbent, with a focus on investigating the influence of the synthesis process conditions (i.e. different coating pH and thermal pre-treatment of raw bauxite at different temperatures prior to the Al coating), on the fluoride adsorption properties of the produced GACB.

Similar to the production of aluminum oxide coated pumice (AOCP), the hard soft acid base (HSAB) concept was explored for the synthesis of GACB. Thus a working hypothesis was that, coating of raw bauxite particle surfaces with Al can make it a better fluoride adsorbent, and that, the higher the amount of Al coated onto the raw bauxite base material, the better would be the fluoride removal capability of the synthesized adsorbent material.

Batch adsorption experiments were used to assess the fluoride adsorption potential of the GACB produced under different process conditions. The effects of major competing co-anions present in natural groundwater, on the defluoridation performance of GACB, was also studied.

The results of the study revealed that surface modification of raw bauxite by an Al coating process, could produce an adsorbent media capable of treating fluoride-contaminated model water to attain the WHO guideline value for fluoride. Thermal pre-treatment of the raw bauxite base material prior to the Al coating resulted in a strong increase in the specific surface area and pore volume of GACB, which contributed to a significant increase in its fluoride adsorption efficiency, compared to that of GACB produced using untreated raw bauxite. The study therefore revealed that the textural properties (i.e. specific surface area and pore volume) of the starting/base material could play an important role in the fluoride adsorption efficiency of the produced adsorbent. The optimal process conditions for the synthesis of GACB comprised of a coating pH of 2, with a thermal pre-treatment of raw bauxite at a temperature of 500°C, prior to the Al coating. It was, however, concluded that even though the thermal-pretreatment of raw bauxite prior to Al coating, in general, significantly enhanced fluoride removal efficiencies of the produced GACB, the efficiencies of GACB synthesized from thermally pre-treated raw bauxite at 200°C and that at 500°C were similar. Therefore thermal pre-treatment at 200°C was recommended, as that could result in a lowered energy cost of production.

The study further revealed that hard surface active sites created by the Al coating procedure, appear to be more effective for fluoride binding compared to hard surface sites of the intrinsic Al content of raw bauxite. This was most likely due to differences in reactivity of the Al incorporated into GACB in the form of gibbsite by the coating process, and that of the natural gibbsite in raw bauxite. The observation of the study suggested that the crystal structure of gibbsite incorporated by the Al coating procedure presumably possessed a higher percentage of specific surface area of edge faces, hence more reactive sites for anion adsorption than the intrinsic gibbsite. The coating procedure was therefore found to be a useful approach to enhance fluoride uptake properties of bauxite and/or similar locally available materials for water defluoridation.

Similar to AOCP, the fluoride adsorption capacity of GACB, based on batch experiment, was also found to be either comparable or higher than that of some reported fluoride adsorbents. It was therefore concluded from this study, that GACB is also a promising defluoridation material that could possibly be used for treating fluoride-contaminated groundwater for rural water supply. It was further revealed that, sulphate and bicarbonate, if present at high concentrations in groundwater, could retard the fluoride adsorption capacity of GACB, while nitrate, chloride and phosphate at concentrations commonly found in groundwater would either have no or negligible effect on the performance of GACB.

The information in this study may be useful for practitioners in search of alternative fluoride adsorbents, with a possibility of a lowered cost of production and sustainability, especially in fluoritic areas where raw bauxite is locally available and/or more accessible. The results from the co-ion study would also be useful for design purposes for field application, as this needs to be considered when GACB is to be used in fluoritic areas with high levels of sulphate in the groundwater. The fluoride removal performance of GACB was, however, found to be reduced after 8 months of storage. Even though the cause of the reduction in adsorption efficiency was not investigated, it may be useful information for water service providers/engineers interested in the use of GACB as water defluoridation material, to either avoid long term storage or investigate further in order to avoid this phenomenon.

7.6 Groundwater defluoridation using aluminol (Al-OH) fuctionalized wood charcoal: Effect of wood source, particle size and field assessment

The study results and related conclusions presented in previous sections, has so far revealed that aluminium modified indigenous materials (AOCP and GACB) are promising fluoride adsorbents. The defluoridation performance of the aluminium modified media, however, needs to be increased in terms of number of bed volumes (BV) of water treated before breakthrough in order to improve the adsorbents practical and economic viability, and hence sustainability when in use. As discussed in section 7.5, a base material with good textural properties (i.e. high specific surface area and pore volume) may be one of the important factors to be considered when selecting a base material for synthesizing aluminium modified adsorbents with enhanced fluoride adsorption properties. The textural properties of virgin pumice and raw bauxite, used as base materials so far, are not optimal, and, the thermal pre-treatment for enhancing these properties could, in practice, contribute to increased costs of the adsorbent production, and also could introduce a need for specialized calcination equipment. Wood charcoal is, however, an alternative indigenous material with naturally good textural properties that can be used as a base material for synthesizing a high-surface-area defluoridation material with enhanced adsorption properties (e.g. capacity and kinetics). Wood charcoal is cheaper, more readily available and accessible in most developing countries, compared to pumice and bauxite. It is also a renewable resource compared with the other base materials, which could contribute to its sustainable use and could also eliminate additional production cost and equipment requirements associated with thermal pre-treatment of base materials.

Even though carbonaceous materials, including wood charcoal, have been known for centuries to possess adsorption capabilities for a wide range of contaminants due to their naturally good textural properties and the presence of surface functional groups, they are also observed to be less effective for adsorption of certain types of contaminants, especially inorganic species present in aqueous solutions including fluoride ions The low fluoride adsorption potential of carbonaceous materials has been ascribed to the non-metallic nature of these materials as compared to metallic materials such as activated alumina (AA), which are known to possess good affinity for aqueous fluoride ions. Based on the chemistry of fluoride, however, the properties of a given carbonaceous material can be tailored through surface treatments, and make it reactive for binding and removal of fluoride ions from solution.

It has also been observed that some laboratory-synthesized fluoride adsorbent materials could work very well under laboratory experimental conditions, but fail to perform under real field conditions when treating natural fluoride-contaminated groundwater. This makes field testing of a laboratory produced material very essential, in the course of developing a potential fluoride adsorbent.

The aims of this component of the research study were therefore to: (i) explore the possibility of providing aluminol (AlOH) functionality to wood charcoal by incorporation of metallic species (Al(III)) into its fabric in order to tailor its surface properties for fluoride adsorption, (ii) investigate the effects of the precursor wood charcoal as well as the adsorbent particle size range on the fluoride removal potential of the aluminol functionalized wood charcoal (AFWC), (iii) assess the performance of AFWC under continuous flow conditions, (iv) investigate the regenerability potential of exhausted AFWC for re-use, (v) screen the efficacies of both AFWC and regenerated AFWC (RAFWC) for water defluoridation under field conditions in pilot- scale columns, using natural fluoride-contaminated groundwater, and, (vi) conduct for comparison, similar pilot-scale column study to determine the adsorption capacity of activated alumina (AA), the industrial standard fluoride adsorbent.

The results of this study showed an increase of the number of bed volumes of water treated (before breakthrough) by the aluminol functionalized wood charcoal (AFWC) by more than 30% compared with that of Al-modified pumice (AOCP), under similar laboratory continues flow conditions. It was therefore concluded that the naturally higher textural properties of locally available wood charcoal makes it a more suitable base material for the synthesis of an aluminum modified fluoride adsorbent, compared with base materials such as bauxite and pumice.

The results further revealed that the source of the plant/tree and, hence wood charcoal used for the production of aluminol functionalized wood charcoal (AFWC), has an influence on its fluoride removal capabilities. The particle size range of AFWC also has an effect on its fluoride adsorption performance, with smaller particle size (0.425 – 0.8 mm) performing about 28% better in terms of number of bed volume of water treated before breakthrough, compared with that of coarser particles (0.8 – 1.12 mm).

The experiments conducted also demonstrated the potential for regenerating fluoride saturated AFWC for reuse, when exhausted. Similar to that of regenerated AOCP (RAOCP), the fluoride removal capacity of regenerated AFWC (RAWC), in terms of the number of bed volumes of water treated before breakthrough was found to increase by about 40 %, after the 1st regeneration cycle. This outcome could contribute to the practical and economic viability of AFWC when in use.

AFWC and regenerated AFWC (RAFWC) were both found capable of reducing fluoride concentration from 4.88 mg/L in natural fluoride-contaminated ground water to ≤ 1.5 mg/L under field conditions. The laboratory fluoride removal performance was therefore reproducible under (real) field conditions.

The grade of activated alumina (AA) tested in the field performed about 3.8 times better than AFWC in terms of number of bed volumes of water treated before breakthrough. The particle size of the AA grade used for the field pilot test was, however, much finer (0.21 – 0.63 mm) than that of AFWC, and this may have contributed to its superior performance over that of AFWC with particle size 0.8 – 1.12 mm, which was available at the time of field test. The AFWC of smaller particle size (0.425 – 0.8 mm) is expected to perform better in the field.

Even though the AA grade exhibited higher performance in the field pilot study, it may not be cost-effective for application in developing countries since it cannot be regenerated and has to be used once and disposed. The frequent disposal due to a one-time use could also pose challenges in terms of the handing and safe disposal to avoid environmental pollution.

It was further concluded based on the field pilot study that, the performances of AFWC and RAFWC were encouraging. It may therefore be too early to conclude on the superiority of the performance of AA over AFWC, since the former (i.e. AA) cannot be regenerated, while AFWC exhibits good potential for regeneration with a possibility of an increasing adsorption capacity with each regeneration cycle. The fluoride adsorption capacity of the freshly produced AFWC, however, require further improvements.

7.7 General outlook, limitations and recommendations

The outcome and conclusions of this research provides improved insight into the occurrence, genesis and distribution of fluoride in the eastern corridor of the Northern of Ghana, which is the most fluoritic area of the region (hence the study area of the research interest). The outcome may be helpful to water providers for strategic planning for the provision of safe drinking water to the populace of the study area in order to prevent the incidence of fluorosis and other fluoride-related health hazards. The methodological approach employed for the study may also be useful to students and researchers interested in conducting similar fluoride occurrence studies in other similar sedimentary rock formations elsewhere.

The research findings and conclusions also provide an insight into the synthesis of fluoride adsorbents by modification of the surface properties of indigenous materials, using an aluminum coating/functionalization process. The use of locally available materials as starting materials for the adsorbent synthesis has the potential to reduce the cost of production as well as improve long term sustainability. The findings also throw some light on how the hard soft acid base (HSAB) concept, could possibly be explored; (i) for the synthesis of innovative adsorbent materials, (ii) the regeneration process of exhausted adsorbents, and (iii) the safe disposal of spent adsorbents into the environment. It appears the HSAB concept, even though a fundamental principle, has so far not been well explored in the search for adsorbent materials.

Thus knowledge obtained from this study may contribute to the search for alternative suitable defluoridation materials that may be sustainable for the treatment of fluoride contaminated-groundwater in developing countries. The extensive use of kinetic and isotherm models conducted in this thesis for the interpretation of adsorption experimental data, could also be helpful to students interested in application of the adsorption technology for water defluoridtaion as well as the removal of other contaminants from the aqueous phase in general.

Even though the research covered many areas concerning fluoride, including the occurrence, genesis and distribution, as wells as the development of fluoride adsorbent material, the regeneration when exhausted, and the safe disposal, there are still many aspects that were either not covered or require further understanding as indicated below, and recommendations for further researchs tudies:

- o Further improvements of the adsorption capacity of the wood charcoal-based fluoride adsorbent need to be explored to enhance the practical and economic viability, and for its possible use in large centralized systems. In this context, the surface chemistry and other relevant properties (e.g. textural properties) of the starting wood charcoal as well the mechanisms involved in the surface modification process using aluminum, require further investigation for an improved understanding.

- o Possibility of modifying the adsorbent syntheses methodology by incorporation of fluoride in the base/precursor material surface treatment process, for enhancing the Al coating/AlOH functionalization process, hence the adsorbent capacity, need to be explored.

o   Given the rather high surface area of AFWC 1 as presented in the thesis, a further assessment of its fluoride adsorption capacity under continuous flow conditions should be re-visited.

o   Since only a 1st cycle of adsorbent regeneration was conducted in this study, further investigation need to be conducted on the feasibility of multiple regeneration.

o   The feasibility (technical and economic) of in-situ regeneration should be studied.

o   An improved scientific explanation as to why the adsorption capacity of the developed adsorbent increases after regeneration is required. The surface chemistries of the freshly synthesized material and the regenerated versions, as well as the mechanisms involved in the regeneration process need to be better understood.

o   The deterioration of the fluoride removal performance GACB after a period of storage also requires further investigation.

o   A detailed/comprehensive assessment of the cost of production of the aluminol functionalized wood charcoal (AFWC) for fluoride adsorption, the regeneration and disposal needs to be conducted, based on a scenario where an optimized production is done in a typical developing country set-up, with locally available wood charcoal.

o   An environmental impact assessment of the production of aluminol functionalized wood charcoal also needs to be conducted, and mitigation measures have to be identified for any potential negative impacts on the environment.

Samenvatting

Eind 2015 werd geschat dat er wereldwijd nog 663 miljoen mensen gebruik maken van onveilige drinkwater bronnen, vooral in de ontwikkelingslanden, inclusief sub Sahara Afrika en Zuid-Azië. Een meerderheid van de getroffen bevolking is arm en leeft op het platteland. Aan het begin van dit onderzoek (2009) werd het deel van de bevolking in de ontwikkelingslanden dat nog geen beschikking had over veilig drinkwater, geschat op 884 miljoen. Hoewel aan het eind van de termijn van de Milleniumsoelstellingen (2015) veel was bereikt, is het duidelijk dat er nog steeds veel moet worden gedaan.

Toegang tot veilig drinkwater is niet alleen van belang voor gezondheid en welzijn van de mens, maar het is ook een fundamenteel grondrecht. De beschikking hebben over veilig drinkwater wordt beschouwd als kritisch en cruciaal voor de algemene ontwikkeling, net als adequate voeding, onderwijs, gendergelijkheid en vooral bestrijding van de armoede in ontwikkelingslanden.

Grondwaterbronnen staan algemeen bekend om hun uitstekende microbiologische en chemische waterkwaliteit, en hebben weinig of geen behandeling nodig voor toepassing als drinkwaterbron. Gebruik van grondwater voor de watervoorziening wordt daarom in het algemeen gerelateerd aan lage investeringskosten en lage exploitatie- en onderhoudskosten. Grondwater is dan ook de meest aantrekkelijke bron voor de drinkwatervoorziening, met name in de sterk verspreide en landelijk gelegen gemeenschappen in ontwikkelingslanden. De chemie van grondwater kan als gevolg van verhoogde concentraties elementen, negatieve effecten hebben op de gezondheid van de gebruikers. In dergelijke situaties vergt de productie van veilig drinkwater een additionele zuiveringsstap. Fluoride is één van de waterkwaliteitsparameters van zorg en aandacht. Veel gronwaterbronnen op veel plaatsen in de wereld overschrijden de WHO norm (1,5 mg/l) en zijn daarmee ongeschikt als drinkwater voor menselijke consumptie als gevolg van schadelijke gezondheidseffecten.

Hoewel een optimale fluoride concentratie (0,5 - 1 mg/l) in drinkwater goed is voor de ontwikkeling van het gebit en de botten, zijn de toxische effecten van een langdurige fluoride inname boven 1,5 mg/l al eeuwen bekend. De gezonheidsrisico's voor mensen bij inname van een overmaat aan fluoriode zijn: fluorose, verandering aan de DNA structuur, verlaging van het IQ bij kinderen, nierproblemen en zelfs de dood bij zeer hoge concentraties (circa 150 - 250 mg/l).

Meer dan 90% van de huishoudelijke waterbehoefte in het noordelijk deel van Ghana (onderzoeksgebied voor deze studie) is afkomstig van grondwaterbronnen. Fluoride verontreiniging in het grondwater in sommige delen van deze regio hebben de bevolking blootgesteld aan fluoride gerelateerde gezondheidsrisico's. Dit heeft er ook toe geleid, dat bronnen die op zich geschikt zijn voor de productie van drinkwater, gesloten moesten worden om fluoride gerelateerde gezonheidsklachten te voorkomen. Het sluiten van deze bronnen veroorzaakt niet alleen economische schade maar belemmert ook de levering van veilig drinkwater aan de bevolking. Als gevolg hiervan is de bevolking genoodzaakt gebruikt te maken van onveilig oppervlaktewater dat in verband wordt gbracht met dodelijke ziekten als cholera en diarree. Toch blijft grondwater de belangrijkste bron voor drinkwater in het noordelijk deel van Ghana, hoewel maar weinig bekend is van de factoren (natuurlijk en/of antropogeen) die van belang zijn voor de grondwaterchemie en die daarmee de bron vormen voor de fluoride verontreiniging en fluoride verspreiding.

Bovendien, als gevolg van het permanante risico mede door een gebrek aan kennis met betrekking tot de effectieve behandeling van fluorose en andere fluoride gerelateerde gezondheidsrisico's, is de verwijdering van fluoride in met fluoride verontreinigd grondwater, als preventieve maatregel, noodzakelijk om inname van een overmaat aan fluoride te voorkomen. Op verschillende plaatsen in de wereld zijn diverse fluoride verwijderingstechnieken ontwikkeld, deze staan bekend als Best Beschikbare Technologiën (BBT's). De huidige methoden hebben zo hun beperkingen woordoor ze in het algemeen niet duurzaam en/of niet geschikt zijn onder de meeste condities, in het bijzonder in afgelegen gebieden in ontwikkelingslanden. Huidige methoden zijn: (i) de Nalgonda techniek, die populair is in enkele Aziatische landen. Deze techniek heeft een beperkte verwijderingsefficiency (tot ongeveer 70%), kent een kritische dosering van chemicaliën en heeft een intensieve monitoring nodig om een effectieve fluoride verwijdering te garanderen. Voorts is het een arbeidsintensief proces en de methode vergt vaardigheden die doorgaans niet beschikbaar zijn in de landelijke gebieden van ontwikkelingslanden; (ii) het contact neerslag proces, dat nog steeds onderwerp van studie is en bovendien lijkt het reactiemechanisme voor de verwijdering van fluoride alleen geschikt bij toepassing van houtskool, als katalysator, dat geproduceerd wordt uit botten. Houtskool geproduceerd uit botten is cultureel niet geaccepteerd in bepaalde gemeenschappen als gevolg van plaatstelijke taboes en uit geloofsovertuiging; (iii) adsorptie door geactiveerd aluminium als adsorberend medium, dit is overigens kostbaar en dan vooral bij topepassing in de ontwikkelingslanden; (iv) adsorptie aan houtskool geproduceerd uit botten, dat niet acceptabel is op veel plaatsen zoals eerder genoemd, en; (v) omgekeerde osmose (RO), deze laatste techniek kent hoge investerings- en onderhoudskosten en vereist een speciale installatie met geschoolde arbeiders en een energiebron.

Door de negatieve gezondheideffecten, als gevolg van de overmaat aan fluoride in drinkwater, blijft de zoektocht naar geschikte technologiën voor de verwijdering van fluoride uit grondwater zeer belangrijk. Van de beschikbare fluoride verwijderingtechnieken wordt het adsorptieproces in het algemeen als het meest geschikt beschouwd en dan in het bijzonder voor toepassing in kleine gemeenschappen. Dit heeft te maken met de vele voordelen als flexibileit, eenvoud van het ontwerp, eenvoudige bediening, kosteneffectiviteit, en de mate van geschiktheid en efficientie voor de verwijdering van fluoride ook bij lage concentraties. De geschiktheid van deze adsorptietechnologie is echter in belangrijke mate afhankelijk van de beschikbaarheid van een geschikt adsorbens. Verschillende adsorberende materialen voor de verwijdering van fluoride zijn ontwikkeld en uitgetest, vooral in laboratoria, zoals, mangaanoxide gecoat aluminium, houtskool uit botten, gebakken klei chips, vliegas, calciet, met natrium geladen montimorilloniet-Na^+, keramische adsorbentia, lateriet, ongemodificeerd puimsteen, bauxiet, zeolieten, vloeispaat, ijzeroxide gecoat zand, gactiveerd quarts en geactiveerd kool. Hoewel sommige van deze adsorbentia enige fluoride adsorptiecapaciteit vertonen, is de toepassing van de meeste materialen doorgaans beperkt door een gebrek aan sociale acceptatie, omdat het niet regenereerbaar is en dus duur of alleen werkzaam is onder extreme pH condities. Dit vereist een pH-correctie en betekent dus additionele kosten, waardoor geschiktheid van een dergelijke fluoride verwijderingsmethode met name in de landelijke gebieden van ontwikkelingslanden zeer beperkt is. Sommige van de bestudeerde fluoride verwijderingsmaterialen zijn beschikbaar in de vorm van deeltjes of poeders, met het risico van verstoppen en het optreden van hydraulische weerstanden, indien ze worden toegepast in vast bed adsorptiesystemen. De zoektocht naar geschikte alternatieven voor de verwijdering van fluoride blijft daarom van belang.

Het hoofdoel van deze studie was daarom tweeledig: (i) bestudering van de grondwaterchemie in het noordelijk deel van Ghana, met de focus op het vóórkomen, het ontstaan en de verspreiding van met fluoride verontreinigd grondwater in de oostelijke corriodor van deze regio (wat het meest fluoride houdende deel is), en (ii) een bijdrage leveren aan de zoektocht naar een geschikte en duurzame fluoride verwijderingstechnologie voor fluoride houdend grondwater voor de productie van drinkwater in ontwikkelingslanden.

Om het eerste onderzoeksdoel te realiseren, werd het klimaat, de geologie, de mineralogie en de hydrologie bestudeerd in een bureaustudie. De chemische gegevens van grondwatermonsters van driehonderdenzevenenvijftig (357) bronnen zijn geanalyseerd met standaard methoden. Univariant statistiche analyse, Pearson's correlatie en principal component analysis (PCA) van de chemische gegevens, werden met het SPSS statistisch softwarepakket, de Piper grafische indeling op basis van de GW tabel software en door thermodynamische berekeningen met PHREEQC, gebruikt als complemantaire methoden om inzicht te verkrijgen in de chemische grondwater samenstelling en om de dominante mechanismen te begrijpen die het vóórkomen van hoge fluoride concentraties beïnlvoeden in het bestudeerde gebied. De chemische gegevens van de grondwatermonsters werd verder geanalyseerd met ArcGIS software om de ruimtelijke verdeling van fluoride te bepalen van de bemonsterde punten in het studiegebied. 'Inverse distance weighting interpolation' (IDW) (met gebruik making van ArcGIS) werd gebruikt om de fluoride verdeling vast te stellen in de niet bemonsterde punten in het bestudeerde gebied om zo de grondwaterfluorideconcentratie niveaus te kunnen voorspellen.

De fluoride concentraties van de 357 grondwatermonsters uit het gebied lagen tussen 0,0 en 11,6 mg/l, met een gemiddelde waarde van 1,13 mg/l. Een relatief hoog percentage (23%) van de monsters overschreed 1,5 mg/l, de WHO norm voor drinkwater. Gebaseerd op de Piper grafische indeling, werden zes grondwatertypen onderscheiden; Ca-Mg-HCO₃,

Ca-Mg-SO$_4$, Na-Cl, Na-SO$_4$, Na-HCO$_3$ en een gemengd water type. De uitgevoerde PCA op de chemische gegevens van het grondwater konden met 4 principal componenten (PC's), 72% van de gegevensvariantie verklaren. De PC's die de overheersende geochemische processen van de grondwaterchemie in het bestudeerde gebied beïnvloeden, kunnen worden verklaart uit de volgende processen: minerale oplossingsreacties, ion uitwisseling en verdamping. Uit PHREEQC berekeningen voor saturatie indices van de grondwatermonsters is gebleken dat deze waren ovderverzadigd voor calciet en onderverzadigd voor fluoriet, wat suggereert dat fluoride in oplossing gaat in gebieden waar dit mineraal aanwezig is. Koppeling van principal component analysis (PCA) aan de gegevens van de evaluatie van de evenwichtstoestand van de grondwatermonsters, gebaseerd op de satuatie indices, suggereert dat de processen die verantwoordelijke zijn voor de 'overall' grondwaterchemie in dat gebied, eveneens van invloed zijn op de verrijking van het water met fluoride. Fluoride rijk grondwater werd voornamenlijk aangetroffen in het Saboba en Cheiponi district en ook in het Yendi, Nanumba Noord en Zuid district. Deze gebieden hebben een onderlaag afkomstig van de Obossom en Oti beddingen, voornamelijk bestaande uit zandsteen, kalksteen, conglomeraat, schalie, arkose en leisteen. De uitgevoerde hydrochemische analyse heeft aangetoond dat, met uitzondering van de bronnen met hoge fluoride concentraties (hoger dan 1,5 mg/l) het grondwater in het bestudeerde gebied geschikt is voor huidhoudelijk gbruik.

De GIS analyse van grondwater gegevens resulteerde in een kaart met de ruimtelijke verdeling van fluoride concentraties in de bemonsterde monsterpunten in het gebied en een voorspellingskaart die behulpzaam kan zijn bij het bepalen van fluoride concentratie niveaus in de niet bemonsterde delen. De verkregen informatie kan verder een hulpmiddel zijn bij de strategische planning voor de levering van grondwater aan de bevolking uit optimale bronnen, zowel wat betreft te hoge fluoriode concentraties als ook ten aanzien van bronnen met een te lage fluoride concentratie (ter voorkoming van cariës).

Het tweede onderzoeksdoel werd bereikt door een combinatie van studies: een synthese van alternatieve fluoride adsorbentia op laboratorium schaal, inclusief een beoordeling van de fluoride adsorberende eigenschappen (op basis van kinetiek en evenwichtscapaciteit), een studie naar de mechanismen verantwoordelijk voor het fluoride verwijderingsprincipe en een vergelijking van zowel geactiveerde aluminium (AA), een indudstriele fluoride adsorbens en verder andere in de lieterauur genoemde fluoride adsorberende adsorbentia. Het onderzoek omvat ook het regeneren van gesynthetiseerd verzadigd materiaal, de economische en praktische haalbaarheid, een veilige verwijdering van verbruikte adsorbentia en een veldonderzoek naar de capaciteit van de gesynthetiseerde adsorbentia om fluoride houdend grondwater te behandelen.

De synthese van fluoride adsorbentia was uitgevoerd om te onderzoeken of, en zo ja op welke wijze de physisch-chemische eigenschappen van lokaal beschikbare materialen te modificeren zijn, zodat dit kan bijdragen aan de lage kosten en duurzaamheid bij toepassing in ontwikkelingslanden. Drie inheemse materialen, te weten: puimsteen, bauxiet en houtskool, die gemakkelijk beschikbaar zijn in veel ontwikkelingslanden, werden gebruikt als grondstof voor modificatie van het coatingsproces. Het gebruik van de verschillende inheemse materialen voor het coatingsproces was bedoeld om na te gaan wat de invloed is van deze uitgangsmateriaal op de fluoride verwijdering en verder om na te gaan wat voor precursor geschikt is voor modificatie. Coating van de inheemse materialen werd gerealiseerd door toepassing van een Al coating/aluminiumhydroxide (AlOH), met 0,5 M $Al_2(SO)_3$ als Al-houdende oplossing en met gebruikmaking van het 'sterk/zwak, zuur/base' (HSAB) concept. Door zijn kenmerken wordt Al (III) geclassificeerd als sterk zuur en F⁻ als sterke base. Op basis van het HSAB concept heeft Al (III) dus een sterke affinitiet voor fluoride, vandaar de keuze voor Al (III) om de inheemse materialen te coaten ten behoeve van de adorptie van fluoride. De drie geproduceerde Al gemodificeerde adsorbentia zijn: Al gecoat puimsteen (AOCP), granulair Al gecoat Bauxiet (GACB) en met Aluminiumhydroxide gecoat houtskool (AFWC). Voor Bauxiet werd de invloed van de synthese omstandigheden op de fluoride adsorberende eigenschappen van het geproduceerde adsorptieproduct (d.w.z. GACB) onderzocht, met als doel de optimalisatie van het syntheseproces. Deze condities omvatten verschillende pH waarden waarbij de coating plaats vond en verschillende proces temperaturen voor thermische voorbehandeling van het Bauxiet, dat voorafgaat aan het coatingproces. Voor houtskool werd het effect van verschillende houtsoorten (4 verschillende bronnen) op de fluorideverwijdering onderzocht. Verschillende karakterisatietechnieken werden gebruikt voor het bestuderen van de verschillende fysisch-chemische eigenschappen van de gesynthetiseerde fluoride adsorptiemiddelen. Deze technieken omvatten: röntgen diffractie (XRD), röntgen fluorescentie (XFR), scanning electronenmicroscopie (SEM), Fourier transformatie infrarood analyse (FTIR), Raman spectroscopie, energie dipersieve röntgenanalyse (EDX), N_2 gasadsorptie/gasdesorptie, ladingstitratie en potentiometrische titratie. Deze infomatie verkregen uit deze karakterisatie technieken was van belang bij het optimaliseren van het coatingsproces van de fluoride adsorberende materialen. Wanneer de geproduceerde adsorptiematerialen waren verzadigd met fluoride vond regeneratie plaats met een Al coating / Aluminiumhydroxide oplossing.

Op batch- laboratorium- en kolomschaal zijn fluoride adsorptie experimenten uitgevoerd om de fluoride verwijderingsefficiency van de geproduceerde adsorbentia te bepalen, maar ook om de kinetische eigenschappen en de fluoride adsorptiecapaciteit vast te stellen. Verschillende kinetische modellen werden gebruikt voor de interpretatie van de kinetische gegevens, waaronder pseudo eerste orde en pseudo tweede orde kinetische modellen; Banghams vergelijking, Elovich en de Weber en Morris intra deeltjes diffusie modellen. Het gebruik van verschilende modellen voor de data interpretatie is nuttig, om dat ze complementair zijn en daardoor leiden tot een beter begrip van de eigenschappen van de gesynthetiseerde adsorptiemiddelen en de aard van het adsorptieproces zelf. Ook de interpretatie van de evenwichtsgegevens werd uitgevoerd met verschillende isotherm modellen, zoals: Langmuir, Freundlich, Temkin, Elovich, BET, gegeneraliseerde Dubinin-Raduskevich en Redilich-Perterson vergelijkigen. De datagegevens uit de kolomexperimenten, in de vorm van doorbraakcurves, werd enook geëvalueerd op basis van drie bekende modellen: de Thomas, de Adams-Bohart en het bed-diepte bepalingsmodel. Vanwege de inherente fouten geassocieerd aan linearisatie, werden zowel geoptimaliseerde lineaire en non lineaire regressie technieken (inclusief floutenanalyse) toegepast voor de bepaling van de 'best fit model' en de overige model gerelateerde parameters. Als gevolg van de complexiteit van de adsorptieprocessen, de kinetiek en de evenwichtsmodellering, werden deze technieken complementair gebruikt in combinatie met Raman spectroscopie, FTIR spectroscopie, meting van de pH_{pzc} en thermische berekeningen, om zo inzicht te krijgen in de mechanismen die betrokken zijn bij de verwijdering van fluoride door de geproduceerde adsorbentia. De effecten van pH en/of co-ionen (sulfaat, chloride, bicarbonaat, nitraat en fosfaat), en een lange bewaartijd op de fluoride verwijderingscapaciteit van de adsorbentia werd vastgesteld op basis van batch adsorptie experimenten. Het effect van deeltjesgrootte op Al gemodificeerd houtskool is onderzocht in een kolomexperiment bij een constant debiet.

Voor alle inheemse materialen (puimsteen, bauxiet en houtskool), bleek modificatie van het deeltjesoppervlak met een Al coating / Aluminiumhydroxide, effectief in het creëren van actieve plaatsen voor fluoide adsorptie in waterige oplossingen, in overeenstemming met het 'sterk en zwak, zuur-base' (HSAB) concept. GACB, AOCP and AFWC waren allen in staat de fluoride concentratie te verlagen van $5,0 \pm 0,2$ tot $\leq 1,5$ mg/l (WHO norm) binnen respectievelijk 32, 1 and 0,5 uur, hetgeen aangeeft dat gemoodificeerd houtskool (AFWC) een superieure kintetiek bezit om fluoride te verwijderen en bovendien was AFWC efficiënter. De thermische behandeling, voorafgaand aan de aluminium caoting droeg significant bij aan de verbetering van de textuureigenschappen (d.w.z. specifiek oppervlak en porie volume). Hierdoor was de effectiviteit van het coatingproces en daardoor ook de fluoride verwijderingscapaciteit van het geproduceerde GACB groter, in vergelijking tot Al coating van onbehandeld bauxiet. De optimale coating pH was 2 en de optimale thermische voorbehandelingstemperatuur was 500°C.

De bron van het houtskool bleek ook van invloed op het fluoride verwijderend vermogen van het geproduceerde AFWC. Op basis hiervan werd de meest geschikte houtskool gekozen, die als basis diende voor de verdere optimalisatie van het Al coatingsproces.

De kinetische- en evenwichtsgegevens van de fluoride adsorptie experimenten van alle gesynthetiseerde adsorbentia zijn redelijk in overeenstemming met de gegevens verkregen uit de kinetische- en evenwichtsmodellen. Op basis hiervan zijn de kinetische en isotherm model parameters geschat, hetgeen nuttig is voor ontwerpdoeleinden. De kinetische en isotherm analyses, de meting van pH_{pzc} en de thermodynamische berekeningen, als mede de FTIR en Raman spectroscopie, suggeren allen dat het mechanisme van fluoride verwijdering complex is en dat het zowel fysische- als ook chmische adsorptieprocessen omvat.

Gebaseerd adsorptie kinetiek experimenten onder laboratoriumcondities is gebleken dat bij een neurtale pH van 7,0 ± 0,2, geschikt voor behandeling van grondwater, de fluoride adsorptie door AOCP beduidend sneller is, dan die van een industriële standaard adsorbens, geactiveerd aluminium (AA). Beiden hadden een vergelijkbare deeltjesgrootte (0,8 - 1,12 mm). Bovendien waren de fluoride adsorptiecapaciteiten van alle Al gemodificeerde adsorbentia, gebaseerd op de Langmuir vergelijkingsisotherm en geschat uit batch adsorptie evenwichts experimenten, vergelijkbaar of hoger dan andere gerapporteerde fluoride adsorbentia (adsorptie bij pH 7,0 ± 0,2). De drie ontwikkelde adsorbentia zijn derhalve veelbelovend en kunnen worden gebruikt voor toepassing bij de behandeling van met fluoride verontreinigd grondwater. De toegepaste procedures uit dit onderzoek kunnen een nuttige aanpak zijn voor de synthese van effective adsorbentia voor gebruik in fluoride houdende grondwater gebieden in ontwikkelingslanden. Hierbij is een besparing op de productiekosten mogelijk, in het bijzonder wanneer de inheemse basimaterialen (puimsteen, bauxiet en houtskool) lokaal beschikbaar zijn. Gebruik van lokaal beschikbare basismaterialen draagt ook bij aan de duurzaamheid.

Zowel AOCP als ook AFWC vertoonden een effecftive fluoride adsorptie in de pH range van 6 tot 9, wat voorkómt dat een pH correctie nodig is, waardoor additionle kosten en operationele problemen kunnen worden vermeden, in het bijzonder bij gebruik in afgelegen gebieden in ontwikkelingslanden. De aanwezigheid van nitraat, bicarbonaat, chloride en fosfaat, in concentraties die doorgaans kunnen worden aangetroffen in grondwater, heeft geen of slechts een verwaarloosbaar effect op de fluoride verwijdering bij toepassing van AFWC en GACB. Sulfaat veroonde had echter wel een verlagend effect op de fluoride verwijdering van AFWC en GACB, ofschoon het effect iets geringer was op de preststies van AFWC.

In een vervolg op de vaststelling van de effectiviteit van AOCP op laboratoriumschaal, zijn kolomexperimenten uitgevoerd, die een meer realistische benadering zijn voor het verkrijgen ontwerpparameters voor praktijkinstallaties. Op basis van deze kolomexperimenten bleek het adsorbent een vergelijkbare fluoride verwijderingscapaciteit te vertonen, waarbij de fluoride concentratie in modelwater werd verlaagd van 5,0 ± 0,2 mg/l naar ≤1.5 mg/l. Het AOCP had een deeltjesgrootte van 0,8 - 1,12 mm, en er was sprake van 165 bed volume (BV) behandeld model water, op het moment van doorbraak. De experimentele doorbraakcurve kon goed worden beschreven met de Thomas, Adams-Bohart en de bed-diepte bepalingstijd (BDST) modellen. De verkregen model parameters zijn nuttig voor het opschalen en ontwerpen van AOCP fluoride verwijderingsfilters, zonder dat aanvullende experimenten nodig zijn. Regeneratie van verzadigd AOCP is haalbaar. De adsorptiecapaciteit van AOCP was na de 1e regeneratiecyclus niet alleen volledig (100%) hersteld, maar was ook 30% hoger, wat een indicatie is van de effectiviteit en het nut van de regeneratieprocedure. De regenereerbaareheid van met fluoride verzadigsd AOCP kan zo op economisch wijze een bijdrage leveren aan de praktiche toepassingsmogelijkheden. In een vergelijkbare test met AFWC (deeltjesgrootte 0,8 - 1,12 mm) was op basis van kolomexperimenten, het aantal behandelde bedvolumes 219 voordat er sprake was van een doorbraak. Dit toont aan dat AFWC een grotere fluoride adsorptiecapciteit heeft (een toename van ongeveer 30%), onder kolomcondities. In het vervolg van het onderzoek werd daarom AFWC gebruikt. Een verlaging van de deeltjesgrootte tot 0,425 - 0,8 mm resulteerde in een verdere optimalisatie van de prestaties, waarbij 282 bedvolumes konden woden behandeld voordat er sprake was van een doorbraak. Dit betekent een verdere toename van 28%.

Bij Granulair aluminum gecoat bauxiet (GACB) werd echter vastgesteld dat de fluoride verwijderingscapaciteit na een opslagperiode van 8 maanden afnam, terwijl de verwijderingsprestaties van AFWC en AOCP vergelijkbaar bleven na een opslagperiode van respectievelijk 8 en 12 maanden. Ook vereist een verbeterde prestatie van GACB een thermische voorbehandeling van het bauxiet basismateriaal, voorafgaand aan het Al coating proces. Hiervoor is een energiebron nodig en speciale calcinatie apparatuur, die beiden kostenverhogend zijn. De prestaties van GACB werden daarom niet verder onderzocht in dit onderzoek.

Omdat AFWC superieure prestaties vertoonde in vergelijking tot AOCP, onder vergelijkbare experimentele condities, werd de werkzaamheid en de prestatie onderzocht in een veld pilot studie in Bongo stad in een gebied in Ghana met hoge fluoride concentraties in het grondwater. Tijdens deze veld experimenten was AFWC (deeltjesgrootte 0.8 - 1.12 mm) beschikbaar. Dit AFWC was in staat de fluoride concentratie van 4.88 mg/l, in natuurlijk grondwater onder praktijk condities, te verlagen tot waarden ≤ 1.5 mg/L. Vergelijkbaar aan de laboratoriumcondities kon een gelijk aantal bedvolumes grondwater worden behandeld, totdat er sprake was van fluoride doorbraak (208,3). De fluoride verwijdering in het laboratorium was dus reproduceerbaar onder praktijkomstandigheden. AFWC kon ook geregeneerd worden nadat het verzadigd was. In een vergelijkbare praktijktest werd de prestatie van geregenereerd AFWC (RAFWC) vastgesteld en de adsorptiecapaciteit nam toe met 40% ten opzichte van vers geproduceerd AFWC (behandeling van 295 bedvolumes, voordat er sprake was was een doorbraak). De trend van de verwijderingsprestsartie was dus vergleijkbaar aan die van AOCP en RAOCP.

Een vergelijking van de fluoride verwijderingprestaties onder vergelijkbare condities tussen AFWC met een deeltjesgrootte van 0,8 - 1,12 mm en geactiveerd aluminium (AA), een industriële fluoride adsorbens, toonde aan dat AA beter presteerde. De prestaties van AA waren echter afhankelijk van het gebruikte type adsorbens en de kwaliteit. De AA kwaliteit die eerder was vergeleken met AOCP had een vergelijkbare deeltjesgrott van 0,8 - 1,12 mm en was volgens de leverancier regenereerbaar. De deeltjesgrootte van de AA die in de praktijktest werd gebruikt was echter veel kleiner (0,21 - 0,63) dan de 0,8 - 1,12 mm, wat vermodelijk heeft bijgedragen aan de betere prestaties. Bovendien was de in de praktijktest gebruikte AA kwaliteit (op basis van de email communicatie met de leverancier) effectief, maar niet regeneerbaar. Hierdoor kan het maar één keer gebruikt worden en moet het daarna worden weggegooid wanneer het verzadigd is met fluoride. Op basis van literatuurgegevens kan toepassing van AA voor de verwijdering van fluoride alleen kosteneffectief worden toegepast in ontwikkelingslanden wanneer het meerdere keren geregeneeerd kan worden.

Verder gaf karakterisatie van (afgewerkte) AFWC met behulp van de US EPA toxiciteitskarakterisering procedure (TCLP) aan dat dit materiaal niet gevaarlijk is en veilig gestort kan worden. AA daarentegen vereist een verdere behandeling voordat het veilig kan worden gestort zonder milieuschade te veroorzaken. Dit kan de operationele kosten bij toepassing van AA verhogen.

Hoewel de huidige fluoride verwijderingscapaciteit van AFWC nog verder verbeterd kan worden, zijn de praktijkprestaties van AFWC en RAFWC bemoedigend, en is het te vroeg om te concluderen dat de prestaties van AA superieur zijn boven de prestaties van AFWC, omdat het in de praktijk gebruikte AA niet geregenereerd kan worden. Het gebruikte AFWC is daarentegen wel regenereerbaar. Bovendien kan AFWC in potentie meerdere malen geregenereerd worden waarbij de prestaties hoogstwaarschijnlijk na iedere regeneratiecyclus toenemen. AFWC kan mogelijk veder worden ontwikkeld en geoptimaliseerd en kan daardoor zeer waarschijnlijk bijdragen aan de levering van veilig drinkwater aan een aantal van de 663 miljoen mensen, die nog steeds gebruik moeten maken van onveilige bronnen, met name in de ontwikkelingslanden.

List of Publications and Conference proceedings

Conference proceedings:

- Salifu, A. Performance of Rope pump technology in the Northern Region of Ghana, 1st International Rope Pump policy workshop, Managua, Nicaragua (2001).

- Salifu, A., Petrusevski, B., Ghebremichael, K., Buamah, R., Amy, G. Fluoride occurrence in groundwater in the Northern region of Ghana, IWA Groundwater Specialist Conference, Belgrade, Serbia (2012).

- Salifu, A., Petrusevski, B., Ghebremichael, K., Modestus, L., Buamah, R., Aubry, C., Amy, G.L. Fluoride removal from drinking water using aluminum (hydr)oxide coated pumice, 11th IWA Conference on Small Water & Wastewater Systems and Sludge Management, Harbin,China (2013).

- Salifu, A., Petrusevski, B., Msenyele, L., Gebremichael, K., Buamah, R., Aubry, C., Amy, G., Kenedy, M.D. Groundwater defluoridation using aluminum (hydr)oxide coated pumice: Laboratory-scale column filter studies and adsorbent regeneration, IWA Groundwater Specialist Conference, Belgrade, Serbia (2016).

- Salifu, A., Petrusevski, B., Mwampashi, E.S., Pazi, I. A., Ghebremichael, K., Buamah, R., Aubry, C., Amy, G. L., Kenedy, M.D. Treatment of fluoride-contaminated groundwater using granular aluminium-coated bauxite: The adsorbent synthesis process, kinetics and equilibrium study. 4th IWA International Symposium on Water and Wastewater Technologies in Ancient Civilizations (WWAC). 17-19, September 2016, Coimbra, Portugal.(Accepted for oral presentation):

Publications (peer reviewed journals):

o Salifu, A., Petrusevski, B., Ghebremichael, K., Buamah, R., Amy, G.L. 2012. Multivariate statistical analysis for fluoride occurrence in groundwater in the Northern Region of Ghana. J. of Contaminant Hydro. 140-14, 34-44.

o Salifu, A., Petrusevski, B., Ghebremichael, K., Modestus, L., Buamah, R., Aubry, C., Amy, G.L. 2013. Aluminum (hydr)oxide coated pumice for fluoride removal from drinking water: Synthesis, equilibrium, kinetics and mechanism. Chem. Eng. J. 228, 63 -74.

o Salifu, A., Petrusevski, B., Mwampashi, E.S., Pazi, I.A., Ghebremichael, K., Buamah, R., Aubry, C., Amy, G.L., Kenedy, M. D. 2016. Defluoridation of groundwater using aluminum-coated bauxite: Optimization of synthesis process conditions and equilibrium study. J. of Env. Manag. 181,108 -117.

Publications (peer reviewed journals in progress/under preparation)

o Dynamic behavior of fluoride adsorption onto aluminum (hydr)oxide coated pumice: laboratory-scale column filter studies, modeling, mechanism and adsorbent regeneration (under preparation for submission to *Environmental Science and Technology*)

o Aluminol (Al-OH) fuctionalized wood charcoal for treatment of fluoride-contaminated groundwater: Effect of wood source, particle size, surface acidity-basicity and field assessment (under preparation for submission to *Carbon Journal*).

o Equilibrium modeling of fluoride adsorption onto aluminum (hydr)oxide coated charcoal: A surface complexation approach (To be prepared for *The Journal of Colloid and Interface Science*).

Biography

Abdulai Salifu is a Civil and Environmental Engineer, and obtained his BSc. degree (Civil Engineering) from the Kwame Nkrumah University of Science and Technology (KNUST), Kumasi, Ghana. He obtained his Post-Graduate diploma degree (passed with distinction) in Community Water Supply and Sanitation, from Loughborough University of Technology (LUT), England (U.K) and his MSc. degree in Environmental Engineering from the University of Newcastle upon Tyne, England (U.K). He worked for several years in the water sector in Ghana as a Water and Sanitation Engineer. He worked in a multi-disciplinary team of professionals comprising of; Hydrogeologist, Sociologists, Health and Hygiene Education Specialist, IT Specialist and Planners, involved mostly in the provision of water supply and sanitation facilities as well as health/hygiene education to rural communities and small urban towns. He was involved in the planning and implementation of several water and sanitation projects in the Northern region of Ghana, financed by several External Support Agencies (ESAs), including the World Bank, European Union (EU), Agence Francais de Developpment (AFD), UNICEF, Japanese International Corporation Agency (JICA) and the Canadian International Development Agency (CIDA). Groundwater was the main source exploited for the community water supply in the Northern region of Ghana. Excess fluoride in the groundwater in some parts of the region was, however, noticed. After observing the adverse human health hazards, the psychological and social impacts, as well as the negative economic impacts due to fluoride contamination of the ground water, coupled with the lack of an appropriate treatment/defluoridation options, Salifu became very interested in research in the area of groundwater defluoridation for drinking water supply. In order to pursue this interest, he applied for a PhD program at the UNESCO-IHE Institute for Water education, Delft, where he undertook a research on the topic: *"**Fluoride removal from groundwater by adsorption technology**: The occurrence, adsorbent synthesis, regeneration and disposal"*. His current interest is to improve on the adsorption capacity of aluminum oxide coated media as fluoride adsorbent, gain an improved insight into the regeneration process of the exhausted adsorbent, field pilot testing and the practical implementation of the technology in fluoritic regions of developing countries.

For Product Safety Concerns and Information please contact our EU
representative GPSR@taylorandfrancis.com Taylor & Francis Verlag GmbH,
Kaufingerstraße 24, 80331 München, Germany

Printed and bound by CPI Group (UK) Ltd, Croydon, CR0 4YY
08/05/2025
01864378-0001